PROTEOMI
AND ANALYTICAL
CHEMISTRY

PROTEOMIC PROFILING AND ANALYTICAL CHEMISTRY
The Crossroads

SECOND EDITION

Edited by

P. CIBOROWSKI
University of Nebraska Medical Center, Omaha, NE, United States

J. SILBERRING
AGH University of Science and Technology, Krakow, Poland;
Polish Academy of Sciences, Zabrze, Poland

AMSTERDAM • BOSTON • HEIDELBERG • LONDON
NEW YORK • OXFORD • PARIS • SAN DIEGO
SAN FRANCISCO • SINGAPORE • SYDNEY • TOKYO

ELSEVIER

Elsevier
Radarweg 29, PO Box 211, 1000 AE Amsterdam, Netherlands
The Boulevard, Langford Lane, Kidlington, Oxford OX5 1GB, UK
50 Hampshire Street, 5th Floor, Cambridge, MA 02139, USA

Notices
Knowledge and best practice in this field are constantly changing. As new research and experience broaden our understanding, changes in research methods, professional practices, or medical treatment may become necessary.

Practitioners and researchers must always rely on their own experience and knowledge in evaluating and using any information, methods, compounds, or experiments described herein. In using such information or methods they should be mindful of their own safety and the safety of others, including parties for whom they have a professional responsibility.

To the fullest extent of the law, neither the Publisher nor the authors, contributors, or editors, assume any liability for any injury and/or damage to persons or property as a matter of products liability, negligence or otherwise, or from any use or operation of any methods, products, instructions, or ideas contained in the material herein.

British Library Cataloguing in Publication Data
A catalogue record for this book is available from the British Library

Library of Congress Cataloging-in-Publication Data
A catalog record for this book is available from the Library of Congress

ISBN: 978-0-444-63688-1

For information on all Elsevier publications
visit our web site at https://www.elsevier.com/

Working together
to grow libraries in
developing countries

www.elsevier.com • www.bookaid.org

Publisher: John Fedor
Acquisition Editor: Kathryn Morissey
Editorial Project Manager: Amy Clark
Production Project Manager: Paul Prasad Chandramohan
Designer: Greg Harris

Typeset by TNQ Books and Journals

CONTENTS

List of Contributors xiii

Preface xv

Chapter 1 Introduction 1

J. Silberring, P. Ciborowski

1.1. Why Do Analytics Matter? 1

1.2. Expectations: Who and What? 3

1.3. What Is Next and Where Are We Going? 4

Chapter 2 Biomolecules 7

A. Burns, P. Olszowy, P. Ciborowski

2.1. Major Features and Characteristics of Proteins
and Peptides 8

2.2. Hydrophilicity and Hydrophobicity 9

2.3. Effect of Protein Fragmentation 11

2.4. Effect of Posttranslational Modifications 15

2.5. Amino Acid Sequence and Separating Conditions 15

2.6. Cysteine and Methionine: Amino Acids Containing
Sulfur.. 17

2.7. Protein Identification and Characterization........................ 19

2.8. Structure–Function Relationship and Its Significance
in Systems Biology Function.. 20

2.9. Protein Folding and Protein–Protein Interactions 21

2.10. Moonlighting of Proteins 22

2.11. Summary 23

References 24

Chapter 3 General Strategies for Proteomic Sample Preparation......25

P. Suder, P. Novák, V. Havlíček, A. Bodzoń-Kułakowska

3.1. Introduction 26

3.2. Inhibitors of Proteolytic and Other Enzymes 26

3.3. Homogenization 28

3.4. Homogenization and Isolation of Organelles 31

3.5. Crude Protein Extraction 34

3.6. Serum and Cerebrospinal Fluid Protein Extraction 36

3.7. Fractionation Based on Size-Exclusion Filters 37

3.8. Chromatographic Methods of Protein Fractionation 38

3.9. Peptide Purification 40

3.10. Detergents, Lipids and DNA 44

3.11. Summary 48

Acknowledgments 48

References 49

Chapter 4 Protein Extraction and Precipitation 51

P. Novák, V. Havlíček

4.1. Introduction 51

4.2. Focus on Hydrophobic Protein Extraction 52

4.3. The Role of Protein Solvation 53

4.4. Protein Precipitation 55

4.5. Salting Out 55

4.6. Isoelectric Point Precipitation 56

4.7. Organic Solvent-Driven Precipitation 57

4.8. Trichloroacetic Acid Precipitation 60

Acknowledgment 61

References 61

Chapter 5 Online and Offline Sample Fractionation............63

M. Smoluch, P. Mielczarek, A. Drabik, J. Silberring

5.1. INTRODUCTION 65

M. Smoluch

5.2. STRONG CATION EXCHANGE, WEAK CATION EXCHANGE,
CONTINUOUS OR STEP GRADIENT?............................. 66

P. Mielczarek, J. Silberring

5.2.1. Historical Perspective 66

5.2.2. Principle of Ion Exchange Chromatography 66

5.2.3. Common Types of Ion Exchange Chromatography
Stationary Phases 68

5.2.4. Choice of Ion Exchanger (Cation or Anion?) 71

5.2.5. Choice of Strong or Weak Ion Exchanger 73

5.2.6. Buffers in Ion Exchange Chromatography 73

5.2.7. Ion Exchange Chromatography in Proteomic Studies 74

References 76

5.3. PROTEIN AND PEPTIDE SEPARATION BASED
ON ISOELECTRIC POINT 77

A. Drabik, J. Silberring

5.3.1. Principles of Isoelectric Focusing 77

5.3.2. Sample Preparation Prior to Isoelectric Focusing 80

5.3.3. Isoelectric Focusing in Liquid State 82

5.3.4. Immobilized pH Gradient Isoelectric Focusing 83

5.3.5. Capillary Isoelectric Focusing 83

5.3.6. Isoelectric Focusing in Living Organisms 84

5.3.7. Summary 85

References 85

5.4. CAPILLARY COLUMNS FOR PROTEOMIC ANALYSES 86

M. Smoluch, J. Silberring

5.4.1. Introduction 86

5.4.2. Conventional Capillary Columns 87

5.4.3. Monoliths 89

5.4.4. Summary and Conclusions 94

5.4.5. Recent Developments 96

References 98

**Chapter 6 Immunoaffinity Depletion of Highly Abundant
Proteins for Proteomic Sample Preparation** **101**

J. Wiederin, P. Ciborowski

6.1. Introduction 101

6.2. Immunodepletion Techniques 102

6.3. Capacity of Immunodepletion Columns and Other
Devices 104

6.4. Reproducibility 105

6.5. Quality Control of Immunodepletion 106

6.6. Albuminome 107

6.7. Summary 113

References 113

Chapter 7 Gel Electrophoresis **115**

A. Drabik, A. Bodzoń-Kułakowska, J. Silberring

7.1. FUNDAMENTALS OF GEL ELECTROPHORESIS 117

A. Drabik, J. Silberring

7.1.1. Introduction 117

7.1.2. Electrophoresis Conditions 119

7.1.3. Agarose Gel Electrophoresis 119

7.1.4. Sample Preparation 120

7.1.5. Separation Conditions 120

7.1.6. Native Polyacrylamide Gel Electrophoresis 121

7.1.7. Electrophoresis in Denaturing Conditions 123

7.1.8. Sample Preparation Prior to SDS-PAGE 124

7.1.9. Staining Techniques 124

7.1.10. Fluorescent Staining 126

7.1.11. Isotope Labeling 127

7.1.12. Data Storage 127

References 127

7.2. TWO-DIMENSIONAL GEL ELECTROPHORESIS 128

A. Bodzoń-Kułakowska, J. Silberring

7.2.1. Introduction 128

7.2.2. First Dimension of Two-Dimensional Electrophoresis:
The Isoelectric Point 129

7.2.3. Second Dimension of Two-Dimensional Electrophoresis:
Molecular Weight 131

7.2.4. Gel Staining 133

7.2.5. Pros and Cons of Two-Dimensional Gel Electrophoresis 135

7.2.6. Quantitation of Protein Using Two-Dimensional Gels 136

7.2.7. Difference Gel Electrophoresis 139

7.2.8. Fluorescent Dyes Used in Difference Gel Electrophoresis 140

7.2.9. Internal Standard 141

7.2.10. Pros and Cons of Difference Gel Electrophoresis 142

References 142

Chapter 8 Quantitative Measurements in Proteomics: Mass Spectrometry 145

A. Drabik, J. Silberring

8.1. Introduction 146

8.2. Absolute Quantitation 146

8.3. Relative Quantitation in Proteomics 149

8.4. Summary 157

Acknowledgments 158

References 158

Chapter 9 SWATH-MS: Data Acquisition and Analysis 161

K. Frederick, P. Ciborowski

9.1. Introduction 161

9.2. Tandem Mass Spectrometry for Quantitative Proteomics 162

9.3. SWATH-MS Data Acquisition 165

9.4. Overview of SWATH-MS Data Analysis 168

9.5. Summary 171

References 172

Chapter 10 Top-Down Proteomics 175

C. Boone, J. Adamec

10.1. Introduction 175

10.2. Protein Separation Methods 176

10.3. Mass Spectrometry of Intact Proteins 183

10.4. Software for Data Analysis 188

References 190

Chapter 11 Proteomic Database Search and Analytical Quantification for Mass Spectrometry 193

M. Wojtkiewicz, J. Wiederin, P. Ciborowski

11.1. Introduction 194

11.2. Protein Databases 196

11.3. Search Engines 199

11.4. Mass Spectrometry Data Searches: Things to Consider 203

11.5. Post-Database Search Data Processing 205

11.6. Searches for Posttranslational Modifications 207

11.7. Summary 208

References 209

Chapter 12 Design and Statistical Analysis of Mass-Spectrometry-Based Quantitative Proteomics Data 211

F. Yu, F. Qiu, J. Meza

12.1. Introduction 212

12.2. Mass Spectrometry-Based Quantitative Proteomics 213

12.3. Issues and Statistical Consideration on Experimental Design 214

12.4. Data Preprocessing for Statistical Analysis 224

12.5. Statistical Analysis of Protein Expression Data 227

12.6. Summary 234

References 235

Chapter 13 Principles of Analytical Validation 239

J. McMillan

13.1. Introduction 239

13.2. Liquid Chromatographic Methods 240

13.3. Validation of a Liquid Chromatographic Method: Identity, Assay, Impurities 241

13.4. Recovery 244

13.5. Accuracy 244

13.6. Precision 245

13.7. Calibration Curve, Linearity, and Sensitivity 245

13.8. Selectivity and Specificity 246

13.9. Stability 247

13.10. Aberrant Results and Errors in Analyses 247

13.11. Further Development of Methods Validation 249

References 250

Chapter 14 Validation in Proteomics and Regulatory Affairs 253

J. Silberring, M. Wojtkiewicz, P. Ciborowski

14.1. The "Uphill Battle" of Validation 253

14.2. Accuracy and Precision 256

14.3. Experimental Design and Validation 258

14.4. Validation of the Method 259

14.5. Validation of Detection Levels...260

14.6. Validation of Reproducibility and Sample Loss 262

14.7. Validation of Performance of Instruments...............................263

14.8. Bioinformatics: Validation of Output of Proteomic Data..........266

14.9. Cross-Validation of Initial Results 267

14.10. Proteomics and Regulatory Affairs...267

References 269

Index **273**

LIST OF CONTRIBUTORS

J. Adamec
University of Nebraska − Lincoln, Lincoln, NE, United States

A. Bodzoń-Kułakowska
AGH University of Science and Technology, Krakow, Poland

C. Boone
University of Nebraska − Lincoln, Lincoln, NE, United States

A. Burns
University of Nebraska Medical Center, Omaha, NE, United States

P. Ciborowski
University of Nebraska Medical Center, Omaha, NE, United States

A. Drabik
AGH University of Science and Technology, Krakow, Poland

K. Frederick
University of Nebraska Medical Center, Omaha, NE, United States

V. Havlíček
Institute of Microbiology, Academy of Sciences of the Czech Republic, Prague, Czech Republic

J. McMillan
University of Nebraska Medical Center, Omaha, NE, United States

J. Meza
University of Nebraska Medical Center, Omaha, NE, United States

P. Mielczarek
AGH University of Science and Technology, Krakow, Poland

P. Novák
Institute of Microbiology, Academy of Sciences of the Czech Republic, Prague, Czech Republic

P. Olszowy
University of Nebraska Medical Center, Omaha, NE, United States; Polpharma SA Pharmaceutical Works, Starogard Gdański, Poland

F. Qiu
University of Nebraska Medical Center, Omaha, NE, United States

J. Silberring
AGH University of Science and Technology, Krakow, Poland; Polish Academy of Sciences, Zabrze, Poland

M. Smoluch
AGH University of Science and Technology, Krakow, Poland

P. Suder
AGH University of Science and Technology, Krakow, Poland

J. Wiederin
University of Nebraska Medical Center, Omaha, NE, United States

M. Wojtkiewicz
University of Nebraska Medical Center, Omaha, NE, United States

F. Yu
University of Nebraska Medical Center, Omaha, NE, United States

PREFACE

The term "proteomics" was coined in the mid-1990s; however, the history of proteomics dates back to the mid-1950s if we consider the first scientific report on 2-dimensional electrophoresis ("Two-dimensional electrophoresis of serum proteins." Smithies, O. and Poulik MD. *Nature*. 1956, 177(4518):1033. PMID: 13322019). Many laboratories used 1- and 2-dimensional electrophoresis for protein analyses, and even though it was not termed "profiling," it was very similar to what we now use in proteomic research. More recently, soft ionization and development of mass spectrometry sequencing of peptides and even intact proteins, widely opened the possibilities for global protein analysis. Suddenly, we found ourselves in the middle of something that was growing rapidly and extremely attractive to pursue scientifically. Our enthusiasm for proteomics is still growing as we enter new frontiers with the development of analytical instrumentation (mass spectrometers, Ultra High Pressure Liquid Chromatography, instruments for nano-flow, analyses, etc.) and computational capabilities of data analysis. We strongly believe that a holistic approach will reveal much knowledge which is yet not known. We have learned that proteomics is a highly interdisciplinary approach but carries a risk of false-positive results if not properly controlled at the analytical level. Hence we learned that proteomics is still short of many standards and widely accepted quality controls. Such standards and quality control measures will be built because of our collective experience and to some extent based on "trial and error" experiments. The field of proteomics is very dynamic technologically, with new tools for sample preparation, sample analyses and data processing being announced almost every day. Tools that we use today might be easily replaced tomorrow by new and greatly improved ones.

It is not an easy task to prepare yet another book on proteomics but we do hope that the content of this second edition of our book will stimulate readers and their interest in using a proteomics approach with care, for the benefit of expansion of our knowledge. Our book is aimed at those researchers who are looking for a relatively compact guide that can walk them through major points of proteomic studies without great detail for each and every step but with a focus on quality control elements, frequently overlooked during daily work maintaining basic concepts and principles of proteomic studies. Therefore, *Proteomic Profiling and Analytical Chemistry: The Crossroads* is written for an audience at various levels, technologists/technicians, undergraduate and graduate students, post-doctoral fellows, scientists as well as principal investigators, to highlight key points ranging from experimental design and biology of systems in question to analytical requirements and limitations.

We are indebted to all of our colleagues, coworkers, and students for their excellent contributions to this book. This book could not have been prepared without the extensive editorial work of Elsevier. Thank you all for your efforts and also for pushing us to complete materials for printing. As always, we have to say that nobody's perfect and we would be grateful for any comments and suggestions that may lead to the improvement of future editions.

P. Ciborowski and J. Silberring

INTRODUCTION

J. Silberring

AGH University of Science and Technology, Krakow, Poland;
Polish Academy of Sciences, Zabrze, Poland

P. Ciborowski

University of Nebraska Medical Center, Omaha, NE, United States

CHAPTER OUTLINE

1.1 Why Do Analytics Matter?   1
1.2 Expectations: Who and What?   3
1.3 What Is Next and Where Are We Going?   4

1.1 Why Do Analytics Matter?

The sum of the optimal steps in the analytical and proteomic analysis (process) is not equal to the optimal process in its entirety! As much as it is a trivial statement, which most of us accept to be true, it has not been fully appreciated despite having a profound impact on the success of laborious, expensive and, in many instances, lengthy projects, as proteomic studies are multistep tasks involving a variety of methods, each governed by its own strengths and limitations. The concept of a proteomic study can be depicted in many ways. In Fig. 1.1, we intentionally highlighted analytical components/phases because the same rules of analytical chemistry/biochemistry apply to discovery as well as validation experiments. The experimental design will be governed by a set of different rules, which does not include instrumentation but has biology heavily involved.

As can be observed, bioinformatics analyses are not depicted in this model, as it is focused on

Proteomic Profiling and Analytical Chemistry. http://dx.doi.org/10.1016/B978-0-444-63688-1.00001-X

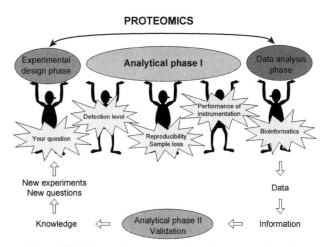

Figure 1.1 Schematic representation of a proteomic study.

analytics. Bioinformatics will be governed by its own set of rules which are applied to the validation of algorithms. Nevertheless, when looking at constituents of a proteomic study, we realized that the scientist conducting such experiments must grasp the overview of not only how biological systems work, but also analytical boundaries for sample preparation, fractionation, and measurements, tools for database searches, statistics, and eventually bioinformatics tools for data analysis. Because of their complexity, proteomic studies should be conducted by a team of experts. As the proteomics field evolves, the collective experience from an increasing number of studies inevitably leads to widely accepted quality criteria. Although significant progress has been made, many questions about uniform quality control criteria remain to be answered. Such answers will result from systematic studies conducted across many laboratories, platforms, and biological systems (models). Therefore, in this book we attempt to highlight in a short, yet comprehensive manner, the impact of the basic principles of analytical chemistry/biochemistry on the final success of a proteomic experiment. We hope that this point of view will help both biologists and chemists to better understand all components of complex proteomic study.

1.2 Expectations: Who and What?

If two scientists, a biologist and chemist, sit at a table and discuss proteomic methodology, they will likely emphasize different aspects of the same study, which in each viewpoint is critical for a successful outcome. Moreover, they quite often speak in technical language that is not fully understood by the other. This is because chemists are focused on sensitivity and accuracy of analytical measurements, while biologists pay attention to explaining biological/pathological effects and are less concerned with exact quantitation of analytes. This resembles the famous poem by John G. Saxe, "The Blind Men and the Elephant," in which everyone tries to identify the part they are touching (ie, biologist/chemist) but nobody can get a sense of the whole system (ie, proteomic study). Biologists are willing to accept a high range of responses, resulting in high standard deviations showing or indicating "trends" in data behavior that support their hypothesis. Chemists, on the other hand, expect data to be expressed by numerical values with high precision, accuracy, reproducibility, and low standard deviation. Indeed, as much as precision of analytical measurements is important, in many instances, such efforts will not improve the overall output discriminating between true and false. This is mostly because, very often, an exact correlation between quantitative change and biological effect is not defined. For example, how important is it to measure a difference between levels of protein expression above 10-fold change when the response of biological system is already saturated by the 5-fold change of this protein? A similar question may arise from enzymology, where the most important factor is enzymatic activity and not the protein expression measured by a typical proteomic approach, which will also measure inactive enzymes. If we bring statisticians and bioinformaticians to the same table as the biologist and chemist, which very often happens, the discussion becomes even more complicated. As illustrated in Fig. 1.2, our question is what do we see on the other side of our office walls when we look for the expertise of our fellow colleagues? It is critical for each of us to peer outside of the walls that confine us and behold the world of those who surround us.

BIOLOGISTS

Chemists and
Mass Spectrometrists

Biostatisticians and
Bioinformaticians

Figure 1.2 What we see on the other side of the wall of our office when we look into the office space of our fellow colleagues with their expertise.

1.3 What Is Next and Where Are We Going?

Since proteomics moved from qualitative to quantitative profiling using liquid-phase-based methods of sample fractionation, it fully entered the domain of analytical chemistry. As much as it is beneficial for proteomics to have a wide range of well-established analytical methods, the complexity of proteomic profiling creates multiple technical issues. First, classical analytical chemistry focuses on high accuracy measurements of single or few compounds at the same time. It allows adjusting methods of sample preparation and analytical parameters with specific objective(s) scarifying measurements of other compounds, which are contaminants, rather than analytes. Importantly, analytical chemistry exploits specific characteristics of analyzed compounds and this concept fulfills its purpose. In contrast, proteomics attempts to measure hundreds and thousands of molecules at the same time which can have a wide range of chemical characteristics (eg, posttranslational modifications of proteins and peptides) and that have a wide dynamic range of concentrations, such as the circumstance with plasma or serum. In the illustration in Fig. 1.1, all steps of a proteomic study are shown as equally important. It would have been a trivial effort if we looked at each step separately. Caveats arise from the connection of these steps as a "well-oiled logically working machine."

In summary, the main goal of this book is to highlight points of junction between proteomics and analytical chemistry, and to link experimental design with analytical measurements, data analysis, and quality control. We also provide a list of points to consider for those who are planning on entering the field of proteomics and have minimal experience.

BIOMOLECULES

A. Burns
University of Nebraska Medical Center, Omaha, NE, United States

P. Olszowy
University of Nebraska Medical Center, Omaha, NE, United States; Polpharma SA Pharmaceutical Works, Starogard Gdański, Poland

P. Ciborowski
University of Nebraska Medical Center, Omaha, NE, United States

CHAPTER OUTLINE

2.1 Major Features and Characteristics of Proteins and Peptides 8
2.2 Hydrophilicity and Hydrophobicity 9
2.3 Effect of Protein Fragmentation 11
2.4 Effect of Posttranslational Modifications 15
2.5 Amino Acid Sequence and Separating Conditions 15
2.6 Cysteine and Methionine: Amino Acids Containing Sulfur 17
2.7 Protein Identification and Characterization 19
2.8 Structure—Function Relationship and Its Significance in Systems Biology Function 20
2.9 Protein Folding and Protein—Protein Interactions 21
2.10 Moonlighting of Proteins 22
2.11 Summary 23
References 24

Proteomic Profiling and Analytical Chemistry. http://dx.doi.org/10.1016/B978-0-444-63688-1.00002-1

2.1 Major Features and Characteristics of Proteins and Peptides

Proteins are very diverse naturally occurring heteropolymers consisting of 20 different monomers (amino acids) in human proteins, varying in length and potentially containing multiple modifications. The physicochemical characteristics of a protein depend on both the overall composition as well as primary sequence of amino acids. Properties of amino acids are grouped based on the functional side chains (R), and one such property is hydrophobicity. If the R group is repelled by water, then it is hydrophobic (nonpolar), eg, valine; whereas hydrophilic (polar) amino acids are attracted to water, eg, arginine. Depending on the primary sequence of amino acids, multiple domains can be contained within the same protein. For example, proteins embedded in the cell membrane have hydrophobic transmembrane domains but hydrophilic extra- and intracellular domains. Another feature of amino acids is the charge of the R group at a neutral pH. However, only the amine and carboxyl termini and the side chains of the following seven amino acids contribute to the net charge of a peptide or protein: tyrosine, cysteine, lysine, arginine, histidine, aspartate, and glutamate. The charge on the amino acids is subject to change based on the pH of the solvent, and the pH of a solvent at which the protein or peptide has a neutral charge is called the isoelectric point (pI). Due to proteins having different chemical properties, the fractionation of proteins, an essential step in any proteomic profiling experiment, is challenging. One approach to accomplish this step is to fragment all the proteins in the sample into short peptides by various chemical and enzymatic methods. The resulting pool of peptides will still form a wide spectrum of molecules ranging from hydrophobic to hydrophilic and acidic to basic; however, each peptide will be easier to separate as a single, narrow peak in liquid chromatography or by isoelectric focusing (IEF).

2.2 Hydrophilicity and Hydrophobicity

Amino acids are hydrophilic or hydrophobic depending on the side chains (R). This feature was used by Jack Kyte and Russell Doolittle, who calculated the hydropathy index [1] based on a measurement of how the R group interacts with water. The calculations are dependent on the free energy of transfer ($\Delta G^{\circ}_{\text{trans}}$) of the solute amino acid between water and condensed vapor phase. A negative $\Delta G^{\circ}_{\text{trans}}$ indicates a strong preference of the R group to water (hydrophilic), whereas a positive value indicates the opposite (hydrophobic).

The hydropathy index model can be applied to predict a protein's tertiary structure. To calculate the hydropathy index for a protein, the individual hydropathy values for an arbitrary number of amino acids, usually 7, 9, 11, or 13 residues, are averaged starting at the N-terminus. These stretches of amino acids, called windows, shift by one amino acid, and the individual hydropathy scores continue to be averaged for each window until the end of the protein is reached. Plotting the hydropathic index versus the position of the amino acid gives a graphical representation of where the hydrophobic and hydrophilic regions of the protein are located. Sharp peaks with positive values in the hydropathy plot correlates well with the hydrophobic regions that will span the membrane.

It is thermodynamically favorable for water to minimize the interaction with the nonpolar hydrophobic moieties. This causes nonpolar molecules to accumulate with each other and form a clathrate structure. A clathrate structure is a cage-like network of water surrounding all the hydrophobic interactions of the nonpolar molecules [2]. Reverse-phase chromatography (RPC) is an important tool that uses hydrophobicity to purify peptides and proteins. RPC involves a nonpolar, stationary phase (C4, C8, or C18, consisting of aliphatic chains containing 4, 8, or 18 carbon atoms, respectively) covalently linked to a solid support and uses a gradient of polar mobile phases to separate chemically different peptides.

The ability of the sample to bind to the stationary phase is proportional to the contact surface area around the nonpolar stationary phase.

For example, a peptide consisting of 7 amino acids has less surface area and hence less hydrophobic amino acids to interact with the stationary phase than a peptide with 16 amino acids. The Kyte–Doolittle analysis [1] determines the hydrophobicity of peptides and the time of elution during chromatography. Keeping with the seven amino acid peptide example, every addition of an amino acid will cause a secondary structure to arise. The secondary structure could diminish the ability of the peptide to bind to the matrix due to the shielding of the hydrophobic R-groups. As the polypeptide chain continues to lengthen, the protein will spontaneously fold to the most thermodynamically stable, tertiary structure by confining the most hydrophobic regions to the interior to minimize the interaction with water [3]. The capacity of an RPC column to purify a peptide is related to the amount of surface area that binds to the nonpolar stationary phase, as mentioned previously. Since a large polypeptide or protein has more surface area and shields the hydrophobic amino acids internally, an RP C18 column that contains more nonpolar hydrocarbons is less efficient in separation. Conversely, smaller peptides need more hydrophobic longer chain lengths to be captured, and therefore usually a C8 or C18 column is used. As much as this property is exploited in the separation of peptides of various lengths, peptides carrying mutations may have quite different physicochemical properties. This effect will depend on the amino acid change, its position and the overall length of the peptide.

In peptide sequencing by mass spectrometry, precursor ions used for consideration must be larger than 600 Da ($m/z > 600^{+1}$ or 300^{+2}). Assuming that the average mass of an amino acid is ~110 Da, peptides to be considered as having a sequence unique for any given protein must consist of five or more amino acids (a.a.). Peptides of such length have limited surface area to interact with the stationary phase used for separation.

2.3 Effect of Protein Fragmentation

As pointed out earlier, protein fragmentation, usually digestion by proteolytic enzyme(s), will generate a set of peptides. Peptides obtained from such cleavage are unique to a single protein and the peptide fingerprint is used for protein identification. This method exploits the specificity of molecular masses from peptide fragments generated by a specific method. As an example, we show in Table 2.1 the characteristics of peptide sets generated by pepsin and trypsin digestion of insulin-like growth factor II (IGF2).

Table 2.1 show the differences of physicochemical properties of peptides derived from the same IGF2 protein fragmented in silico by trypsin and pepsin. It is important to note that trypsin derived a set of peptides that are either acidic with isoelectric points (pIs) below 4.33 or basic with pIs above 10.34. If such a digest is further fractionated based on isoelectric focusing, eg, OFFGEL, we expect the peptides to be on an opposite ends of fractionation spectrum. On the contrary, complete pepsin digestion will only generate four peptides (Table 2.1) suitable for protein identification by mass spectrometry based on the length requirement of peptides (600 Da singly charged). Trypsin digestion will generate four peptides with m/z bigger than 300 for doubly charged species, which will be fragmented for MS/MS identification and, if tagged, also for quantitation. Table 2.1 provides an example of observed peptides generated by pepsin digestion of IGF2. In this case, due to miss-cleavages, pepsin digestion generated nine such peptides. Considering the contribution of each peptide to a high confidence of identification and quantitation, in this particular instance, pepsin digest will have an analytical advantage over tryptic digest. Another issue is that the selection of peptides for a multiple reaction monitoring (MRM) experiment will be reduced when IGF2 is fragmented using trypsin rather than pepsin. Therefore, depending on the focus of the proteomic experiment, choice of proteolytic enzyme or other means of peptide fragmentation may affect the accuracy of quantitation. It also needs to be noted that in silico digestion using

Table 2.1

A. Composition and properties of fragments from *in silico* trypsin digestion of IGF2

Fragment no.	Isoelectric point[a]	Hydrophobicity[b]	Molecular weight (Da)	Amino acid residues	Amino acid sequence[c]
1	4.13	−3.4	2761	24	AYRPSETLCGGE LVDTLQFVCGDR
2	11.01	−6.5	1187	10	GFYFSRPASR
3	10.34	−1.1	360	3	VSR
4	10.34	−4.5	174	1	R
5	10.34	−5.3	261	2	SR
6	4.33	4.6	1055	9	GIVEECCFR
7	4.17	5.8	1699	16	SCDLALLETYCATPAK
8	3.67	−4.3	234	2	SE

B. Composition and properties of fragments from *in silico* pepsin digestion of IGF2

Fragment no.	Isoelectric point[a]	Hydrophobicity[b]	Molecular weight (Da)	Amino acid residues	Amino acid sequence[c]
1	6.38	−10.6	823	7	AYRPSET
2	5.79	3.8	131	1	L
3	3.67	−1.8	364	4	CGGE
4	5.79	3.8	131	1	L
5	3.49	0.0	333	3	VDT
6	5.79	3.8	131	1	L
7	5.79	−3.5	146	1	Q
8	5.95	2.8	165	1	F
9	5.79	−2.1	606	6	VCGDRG
10	5.77	1.5	328	2	FY
11	5.79	2.8	165	1	F
12	10.22	−15.0	2148	19	SRPASRVSRRSRGIVE ECC
13	5.79	2.8	165	1	F
14	5.95	−6.3	480	4	RSCD
15	5.79	3.8	131	1	L
16	5.79	1.8	89	1	A
17	5.79	3.8	131	1	L
18	5.79	3.8	131	1	L
19	4.33	−9.9	1199	11	ETYCATPAKSE

Continued

Table 2.1 *(continued)*

C. Composition and properties of observed fragments from pepsin digestion of IGF2 [14]

Fragment no.	Isoelectric point[a]	Hydrophobicity[b]	Molecular weight (Da)	Sequence	Amino acid sequence[c]
1	6.38	−6.8	936	1-8	AYRPSETL
2 & 7	3.93	0.0	1436	9-13	2: CGGEL
				45-52	7: ECCFRSCD
2A & 7	3.93	3.9	1650	7-13	2A: TLCGGEL
					7: ECCFRSCD
2B & 7	3.93	3.8	1549	8-13	2B: LCGGEL
					7: ECCFRSCD
3	3.49	0.3	575	14-18	VDTLQ
3A	3.49	3.8	446	14-17	VDTL
4 & 10	6.09	−2.1	2030	19-27	4: FVCGDRGFY
				59-67	10: YCATPAKSE
4A & 10	6.09	−2.2	1867	19-26	4A: FVCGDRGF
					10: YCATPAKSE
4 & 10A	4.50	−6.4	2097	57-67	4: FVCGDRGF
					10A: ETYCATPAKSE
5	12.78	−18.9	1619	28-41	FSRPASRVSRRSRG
5A	10.34	−3.1	664	28-33	FSRPAS
6	3.67	5.2	359	42-44	IVE
8	5.79	9.4	315	53-55	LAL
9	3.67	−0.4	361	56-58	LET

[a]Average pI calculated from a computer program based on Kozlowski L. 2007−11 (http://isoelectric.ovh.org/).
[b]Calculated according to Kyte and Dolittle [1].
[c]Single-letter code for amino acids used.

tools such as a peptide cutter (http://web.expasy.org/peptide_cutter/) are beneficial but often do not reflect the real effect of proteolytic digestions.

2.4 Effect of Posttranslational Modifications

Chemical modifications have an impact on the overall chemical properties of proteins and peptides. A single-site acetylation of a 50 kDa or larger protein may not be detectable by many analytical methods. Acetylation is difficult to detect because the increase in hydrophobicity of acetylated versus non-acetylated proteins can be negligible if a protein is, by itself, quite hydrophobic in nature. The situation changes significantly when such a protein is enzymatically digested for proteomic profiling. In this situation, acetylation might be located on a relatively short peptide, eg, 8 to 10 a.a., and have a profound impact on the overall hydrophobicity of this molecule, leading to a shift in elution time in RP-LC. Physicochemical properties of proteins and peptides are further complicated when multiple residues on one protein or a longer peptide are modified, and, in extreme cases, when modifications are heterogeneous.

2.5 Amino Acid Sequence and Separating Conditions

There is no "one size fits all" solution in protein and peptide analyses and chromatographic separation of peptides. Thus the analytical approach will depend on how we match the structure and properties of peptides of interest with characteristics of separation media. This is a very important and quite often neglected issue when considering all details of proteomic profiling experiment(s).

In most instances prepacked columns or bulk resins are used, such as an RP-LC column, without analyzing the type of resin in any given column. It is more evident now than 10 years ago that the success of proteomic profiling requires narrowing the scope of investigation to improve sensitivity and specificity.

One example is immunodepletion of the most abundant proteins from plasma, serum or CSF samples to reduce the dynamic range of protein concentrations. The example provided shows a selectivity comparison between different silica-based media at pH 2.0 and 6.5 (Fig. 2.1) using a mixture of closely related angiotensin peptides. Peptides 1, 2, and 3 are different in one amino acid and their sequences are as follows: (1) RVYVHPI, (2) RVYIHPI and (3) RVYVHPF, respectively. While the

1. Val4-Ile7-AT III (RVYVHPI)
2. Ile7-AT III (RVYIHPI)
3. Val4-AT III (RVYVHPF)
4. Sar1-Leuß-AT II (Sar-RVYIHPL)
   (Sar=sarcosine, N-methylglycine)
5. AT III (RVYIHPF)
6. AT II (DRVYIHPF)
7. des-Asp1-AT I (RVYIHPLFHL)
8. AT I (DRVYIHPFHL)

Columns:
a) and e) Sephasil Protein C4 5 μm 4.6/100
b) and f) Sephasil Peptide C8 5 μm 4.6/100
c) and g) Sephasil Peptide C18 5 μm 4.6/100
d) and h) μRPC C2/C18 ST 4.6/100

Eluent A (pH 2): 0.065% TFA in distilled water
Eluent B (pH 2): 0.05% TFA, 75% acetonitrile
Eluent A (pH 6.5): 10 mM phosphate
Eluent B (pH 6.5): 10 mM phosphate, 75% acetonitrile
Flow: 1 ml/min
System: ÄKTApurifier
Gradient: 5–95% B in 20 column volumes

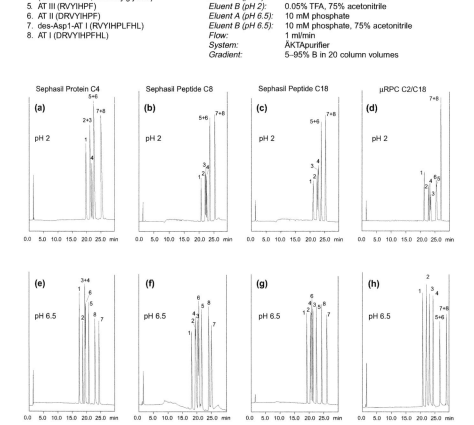

Figure 2.1 Selectivity comparison between different silica-based media at pH 2.0 and 6.5. A mixture of closely related angiotensin peptides was used as sample. Work by Amersham Biosciences AB, Uppsala, Sweden. Reproduced with permission from GE Healthcare, Inc.

third peptide has a distinct value of mean hydrophobicity (0.08) compared to the first and second peptides, 0.32 and 0.37, respectively, peptides (2) and (3) can be coeluted (Fig. 2.1) or eluted separately (Fig. 2.1H). On the other hand, if peptides (1), (2) and (3) are to be eluted separately, peptides (5) and (6) and peptides (7) and (8) will be coeluted (Fig. 2.1). Such a dual factor effect on peptide separation can be exploited with great benefits if the project is focused to address more specific questions than full-range, unbiased proteomic profiling.

Similarly, in gel-based proteomic separation, the application of different conditions such as the percentage of the gel and buffer system used may favor separation of protein in different molecular-weight regions. Application of various conditions is broader for one-dimensional gel electrophoresis (separation based on protein molecular weight) than in two-dimensional gel electrophoresis (separation based on molecular weight and pI).

2.6 Cysteine and Methionine: Amino Acids Containing Sulfur

Cysteine and methionine are two amino acids that contain sulfur. Methionine is an essential amino acid, whereas cysteine is synthesized from methionine and therefore is nonessential. Cysteine is classified as a polar, noncharged amino acid while the side chain of methionine is quite hydrophobic. The hydropathy index of methionine and cysteine is positive and equal to 1.9 and 2.5, respectively, according to the Kyte and Doolittle scale [1]. Unlike cysteine, the sulfur of methionine is not highly nucleophilic, although it will react with some electrophilic centers. Methionine is generally not a participant in the covalent chemistry that occurs in the active centers of enzymes. Thiolate anion is formed after ionization of cysteine in basic solutions and does not change the biophysical character of this amino acid. Therefore, it is uncommon to find cysteine on the surface of a protein even after ionization. The sulfur of methionine and cysteine is subject to oxidation, but cysteine is almost always reduced and blocked by alkylation. Therefore, oxidated methionine but not cysteine is added to database searches of tandem mass spectra. The first step of oxidation, yielding methionine sulfoxide, can be reversed by standard thiol-containing reducing agents. The second step yields methionine sulfone and is effectively irreversible.

When oxidized, cysteine residues can form disulfide bonds, strengthening a protein's tertiary and quaternary structures. Additionally, many metal-containing proteins use cysteines to hold their metals in place, as the sulfhydryl side chain is a strong metal binder. There are a few reasons why sulfur atoms in amino acids do not affect position of those amino acids in proteins. One of the most important aspects is the strong ability to create disulfide bonds in comparison with creation of weaker, noncovalent hydrogen bonds with water. On the other hand, the weaker ability to attract electrons (in comparison to oxygen) results in lack of hydrogen bonds using a sulfur atom.

Cysteine stabilizes the tridimensional structure of proteins, which is critical for extracellular proteins that might be exposed to harsh conditions. Proteins containing multiple disulfide bridges are more resistant to eg, thermal denaturation, and thus may maintain their biological activity at more extreme conditions.

The existence of disulfide bridges inside a protein (intramolecular) and/or between different polypeptide chains (intermolecular) make it necessary to break those bonds before proteomic analysis for making the protein accessible to proteolytic fragmentation. The standard approach is a two-step procedure. In the first step, proteins are reduced using dithiothreitol (DTT $-$ $C_4H_{10}O_2S_2$) or mercaptoethanol, although the latter agent is now used rather seldom. In this step, disulfide bridges break, yielding free sulfhydryl groups. In the second step, free sulfhydryl groups are alkylated to prevent reoxidation and the formation of the bridges.

The biological importance of sulfur-contacting amino acids is multifold. Methionine is necessary for the synthesis of proteins. It forms S-adenosyl-L-methionine (SAM), which serves as a methyl donor in reactions, prevents fatty liver through trans-methylation and choline formation, and can lower toxic acetaldehyde levels in humans after alcohol ingestion. It also plays an important role in preserving the structure of the cell membrane [4] and it has an important function for some reactions involved in protein and DNA synthesis [5]. Cysteine is found in beta-keratin, an important component of skin, hair and nails. A greater number of disulfide bonds causes keratin to be very hard, like in nails or teeth, or flexible, like in hairs. A smaller number of disulfide bonds creates soft keratin in skin. The human body uses cysteine to produce the antioxidant glutathione, as well as the amino acid taurine. The body can also convert cysteine into glucose for a source of energy. Cysteine also plays a role in the communication between immune system cells.

2.7 Protein Identification and Characterization

High-confidence protein identification and in-depth characterization in a proteomic experiment is the most favorable goal. Although new tools have been developed during the last decade, inherent properties of proteins and peptides create limitations on how much information we can obtain. For example, using one enzyme for protein fragmentation might generate peptides that are either too short or too long. For a protein with high confidence identification, two or three peptides are usually sufficient; however, for characterization and/or investigation of specific regions of a protein, it might not be enough. For example, histones are highly posttranslationally modified and contain multiple consecutive lysine residues. It is analytically challenging to identify the exact position of, eg, methylation or acetylation. Therefore, protein characterization usually requires more than one analytical approach, which in consequence will require more biological material not always abundantly available.

2.8 Structure—Function Relationship and Its Significance in Systems Biology Function

The major goal of proteomic profiling experiments is to gain insight into how complex biological system works; therefore the most desirable outcome is new functional information. When proteomics emerged in the mid-1990s everyone was fascinated with the ability to identify (catalog) tens, hundreds, and then thousands of proteins in one analytical experiment. This excitement did not last long, because it became understood that desired information is in relative quantitation rather than the presence or absence of a particular protein. At this point the presence of posttranslational modifications increased the complexity of proteomic experiments by at least two orders of magnitude. New experimental approaches have been proposed and collectively there has been great progress in accumulating huge amounts of data. Although we make significant steps in the biological interpretation of the massive data, our knowledge about how biological systems are functioning grows at a disproportionally low rate. The two hurdles in progress are the correlation of protein structure and function and protein localization and function. The latter phenomenon is also called protein moonlighting. This brings us to question what a protein structure represents in defining its biological function and, further on, how a protein's structure defines its physiological function.

What if we assume that similar sequences of proteins represent similar functions while different sequences are responsible for different functions? We will certainly find many examples to support such an assumption. Let's consider transmembrane domains of receptors, which are hydrophobic and have helical structures to be accommodated by a hydrophobic environment of a lipid bilayer. Further on, integrins alpha 1, 2 and 4 have single transmembrane helical domains, which all play one synonymous function: anchoring these proteins into the cell membrane. They are all close to the C-terminal end of the

Integrin alpha 1

1131 ISKDGLPGRVP**LWVILLSAFAGLLLLMLLILALW**KIGFFKRPLKKKMEK-COOH 1179

Integrin alpha 2

1121 IMKPDEKAEVPT**GVIIGSIIAGILLLLALVAILW**KLGFFKRKYEKMTKNPDEIDETTELSS-

COOH 1181

Integrin alpha 4

971 RPKRYFT**IVIISSSLLLGLIVLLLISYVMWK**AGFFKRQYKSILQEENRRDSWSYINSKSNDD-

COOH 1132

Figure 2.2 Amino acid sequences of transmembrane domains of integrins alpha 1, 2, and 4.

polypeptide chain; however, all of them have a different primary structure (Fig. 2.2).

As we know, integrins are responsible for transmitting signals related to numerous functions and are part of alpha/beta heterodimers.

2.9 Protein Folding and Protein–Protein Interactions

Proteins fold to reach the conformation associated with their function. The process of protein folding is not fully understood; however, we know that most proteins are folded during or right after synthesis. Many proteins, although properly folded, need further processing and help from chaperones to reach their final functional structure. Many proteins are maintained unfolded by chaperones; otherwise they could not be transported outside of the cell. For example, *E. coli* developed a specialized Sec translocase system for posttranslational translocation of proteins [6,7]. This system is a complex of the ATP-driven motor protein SecA and the SecYEG proteins functioning as a membrane-embedded translocation channel. One of the features of this system is that only unfolded proteins can be translocate, and thus they need to maintain the translocation-competent state. SecB holdase, which is an export-dedicated molecular chaperone, prevents proteins to be translocated from folding and aggregating. Summarizing, if we extract all proteins from a cell, denature, and fragment using eg, trypsin and quantitate based on resulting peptides, we are unable to conclude whether the protein was unfolded and complexed with a chaperone and will contribute to the active pool outside of the cell was folded and never destined to be exported. Even if we measure the stoichiometric ratio of chaperone to protein, the evidence of their function and quantitation gives us limited information. Another example of structural complexity is the presence of flexible regions of proteins, which may lead to conformational changes upon self-interactions forming homopolymers or upon interactions with other proteins.

Protein−protein interaction might be mediated by an induced-folding mechanism. This mechanism has been proposed for disabling the intrinsic antiviral cellular defense mechanism by HIV-1 Vif protein [8]. Vif neutralizes two components of a human antiviral defense mechanism, APOBEC3G and APOBEC3F, by engaging them with the cellular protein complex of EloB, EloC, Cul5 and Rbx2 to promote degradation via proteasomal pathway. In this example, participation of Vif in such a complex determines one of its many functions.

2.10 Moonlighting of Proteins

Protein moonlighting is a phenomenon acquired during the evolutionary process when a single protein performs more than one function, which is also associated with specific localization for specific function. This phenomenon was first described by Joram Piatigorsky and Graeme Wistow in late 1980s [9], but gained more attention after given the term "moonlighting" by Constance Jeffery in 1999 [10]. The first proteins shown to moonlight were crystalline and other enzymes [11]; later proteins such as receptors, ion channels, chaperones or structural proteins [12] expanded this list.

Due to the lack of a systematic experimental approach, moonlighting properties of proteins have been found as a result of other studies that did not directly target dual functionality of proteins of interest. Nevertheless, the number of moonlighting proteins is rapidly increasing, indicating that moonlighting proteins appear to be abundant in all kingdoms of life [13]. We may speculate that the list of such proteins is not complete and future studies will add new proteins to list.

The moonlighting phenomenon may also contribute to various diseases. Therefore, while interpreting the results of proteomic studies, and in particular when the objective of such studies is to connect changes in expression levels with function(s) having a biological effect, protein moonlighting needs to be considered. If a protein binds other molecules, whether small molecules, carbohydrates or other proteins, it may acquire new function, which can be also associated with different localization. It has to be determined whether such a property falls under moonlighting or not, and this can be argued both ways. It is important for determination of biological function(s) of investigated proteins. It becomes more complicated when the pool of relatively abundant extracellular protein circulating in body fluid is considered. Proteins circulating as complexes with antibodies might not be properly quantitated using an ELISA assay and MRM-based quantitation after proteolytic fragmentation may give different concentrations. Very often, extracellular proteins are considered as a homogenous population of molecules; in fact they may represent an array of functionally different subsets. It is also possible that only one subset might be relevant as a biomarker, whether diagnostic or reflecting molecular mechanisms of the underlying pathological state.

2.11 Summary

Diverse features of proteins as heteroploymers further augmented by potential posttranslational modification creates enormous challenge for proteomics at the level of analytical analysis as well as data interpretation. Our recommendation is that prior to any proteomic study, one should perform a thorough analysis of physicochemical properties of proteins of interest to match their characteristics with analytical methodology.

References

[1] Kyte J, Doolittle RF. A simple method for displaying the hydropathic character of a protein. J Mol Biol 1982;157(1):105–32.

[2] Biswas KM, DeVido DR, Dorsey JG. Evaluation of methods for measuring amino acid hydrophobicities and interactions. J Chromatogr A 2003;1000(1–2):637–55.

[3] Cserhati T, Szogyi M. Role of hydrophobic and hydrophilic forces in peptide-protein interaction: new advances. Peptides 1995;16(1):165–73.

[4] Vara E, Arias-Diaz J, Villa N, Hernandez J, Garcia C, Ortiz P, et al. Beneficial effect of S-adenosylmethionine during both cold storage and cryopreservation of isolated hepatocytes. Cryobiology 1995;32(5):422–7.

[5] Ahmed HH, El-Aziem SH, Abdel-Wahhab MA. Potential role of cysteine and methionine in the protection against hormonal imbalance and mutagenicity induced by furazolidone in female rats. Toxicology 2008;243(1–2):31–42.

[6] Bechtluft P, Kedrov A, Slotboom DJ, Nouwen N, Tans SJ, Driessen AJ. Tight hydrophobic contacts with the SecB chaperone prevent folding of substrate proteins. Biochemistry 2010;49(11):2380–8.

[7] Driessen AJ, Nouwen N. Protein translocation across the bacterial cytoplasmic membrane. Annu Rev Biochem 2008;77:643–67.

[8] Bergeron JR, Huthoff H, Veselkov DA, Beavil RL, Simpson PJ, Matthews SJ, et al. The SOCS-box of HIV-1 Vif interacts with ElonginBC by induced-folding to recruit its Cul5-containing ubiquitin ligase complex. PLoS Pathog 2010;6(6):e1000925.

[9] Wistow GJ, Piatigorsky J. Lens crystallins: the evolution and expression of proteins for a highly specialized tissue. Annu Rev Biochem 1988;57:479–504.

[10] Jeffery CJ. Moonlighting proteins. Trends Biochem Sci 1999;24(1):8–11.

[11] Chen JW, Dodia C, Feinstein SI, Jain MK, Fisher AB. 1-Cys peroxiredoxin, a bifunctional enzyme with glutathione peroxidase and phospholipase A2 activities. J Biol Chem 2000;275(37):28421–7.

[12] Kourmouli N, Dialynas G, Petraki C, Pyrpasopoulou A, Singh PB, Georgatos SD, et al. Binding of heterochromatin protein 1 to the nuclear envelope is regulated by a soluble form of tubulin. J Biol Chem 2001;276(16):13007–14.

[13] Huberts DH, van der Klei IJ. Moonlighting proteins: an intriguing mode of multitasking. Biochim Biophys Acta 2010;1803(4):520–5.

[14] Rickard EC, Strohl MM, Nielsen RG. Correlation of electrophoretic mobilities from capillary electrophoresis with physicochemical properties of proteins and peptides. Anal Biochem 1991;197(1):197–207.

GENERAL STRATEGIES FOR PROTEOMIC SAMPLE PREPARATION

P. Suder
AGH University of Science and Technology, Krakow, Poland

P. Novák and V. Havlíček
Institute of Microbiology, Academy of Sciences of the Czech Republic, Prague, Czech Republic

A. Bodzoń-Kułakowska
AGH University of Science and Technology, Krakow, Poland

CHAPTER OUTLINE

3.1 Introduction   26
3.2 Inhibitors of Proteolytic and Other Enzymes   26
3.3 Homogenization   28
3.4 Homogenization and Isolation of Organelles   31
3.5 Crude Protein Extraction   34
3.6 Serum and Cerebrospinal Fluid Protein Extraction   36
3.7 Fractionation Based on Size-Exclusion Filters   37
3.8 Chromatographic Methods of Protein Fractionation   38
3.9 Peptide Purification   40
   3.9.1 Phosphopeptides   41
   3.9.2 Glycopeptides   42
3.10 Detergents, Lipids and DNA   44
   3.10.1 Detergents   44
   3.10.2 Lipids and DNA   46
3.11 Summary   48
Acknowledgments   48
References   49

Proteomic Profiling and Analytical Chemistry. http://dx.doi.org/10.1016/B978-0-444-63688-1.00003-3

3.1 Introduction

Technological developments in the field of proteomics clearly indicate that a significant increase of sensitivity, resolution and mass accuracy of mass spectrometers will not be a way to correct or compensate for the issues associated with sample preparation. These issues include initial sample preparation, such as homogenization of tissues, cell lysis, sample clean-up, fractionation, enrichment, as well as all other steps to maintain optimal preparative conditions. The latter is of an increasing importance in the case of high-throughput experiments that include hundreds of samples, almost each in limited supply, eg, clinical material.

There is quite extensive literature in the area of proteomic sample preparation, as well as biotech companies providing protocols and commercial products that allow researchers to rely on their reproducibility and efficiency in designing profiling experiments. For some types of experiments, reference samples, or internal standards are available to help normalize experimental samples at the analytical level. However, we should keep in mind that each experiment is unique and may require minor or major modifications of sample preparation protocols. It is worth remembering that some procedures, still used in laboratories, were created a decade or more ago, when some technologies were not available. Thus, we should adopt or modify older routines to the newer experimental conditions and analytical demands.

Starting from the preparation stage, to receive a proteomic sample suitable for further experiments, we have to accomplish two goals:

1. Create the less complex sample by prefractionation, depletion of the most abundant proteins, purification from DNA, lipids, etc.
2. Clean-up the sample from impurities like salts or remaining solid particles.

3.2 Inhibitors of Proteolytic and Other Enzymes

The major goal of extraction is to release as much of the protein contents as possible, and such a process is not selective. Therefore during extraction, the enzymes such as proteases, phosphatases and others are released. Activity of these enzymes must be inhibited to prevent protein structure changes that may mask the true state of the sample at the time when the biological experiment is terminated and the analytical phase starts. The first step is to place the sample on wet ice to reduce the temperature and slow down enzymatic activity along with addition of a cocktail of inhibitors. In many instances, samples are snap-frozen and a cocktail of inhibitors is added prior to sample thaw and lysis (homogenization), ie, inhibitors are present in an extraction buffer.

Many of protease inhibitors are commercially available alone or in a premixed combination. Each inhibitor is directed to a different class of enzymes. Pepstatin A strongly inhibits acid proteases—pepsin, cathepsin D and renin [1]. Leupeptin inhibits serine and cysteine proteases—plasmin, trypsin, papain, calpain, and cathepsin B; it does not inhibit pepsin, cathepsins A and D, thrombin, or α-chymotrypsin. Antipain inhibits, however reversibly, serine/cysteine proteases and some trypsin-like serine proteases. Its mode of action resembles that of leupeptin, but its plasmin inhibition is less potent than its cathepsin inhibition. Aprotinin inhibition is more than that observed with leupeptin. Aprotinin is a competitive serine protease inhibitor blocking the activity of trypsin, chymotrypsin, kallikrein and plasmin. The inhibitory mechanism of aprotinin is to form stable complexes with the proteases and thus to block the active sites of enzymes. Chymostatin is a strong inhibitor of many proteases, including chymotrypsin, papain, chymotrypsin-like serine proteinases, chymases, and lysosomal cysteine proteinases such as cathepsins A, B, C, B, H, and L. It weakly inhibits human leukocyte elastase. It also inhibits the lysosomal proteinase cathepsin B, and the soluble Ca^{2+}-activated proteinase [2]. Phenylmethanesulfonyl fluoride (PMSF) is a specific trypsin and chymotrypsin inhibitor [3], and 4-(2-aminoethyl)benzenesulfonyl fluoride (AEBSF) has been shown to inhibit trypsin, chymotrypsin, plasmin, kallikrein and thrombin. As an alternative to PMSF, AEBSF offers lower toxicity, improved solubility in water and improved stability in aqueous solution [4]. Ethylenediaminotetraacetic acid (EDTA) is often used to deactivate metalloproteases and other enzymes whose activities are based on the presence of divalent ions. All of these inhibitors are used to prevent a general process of protein degradation; however, targeted proteomic profiling studies may also require using different inhibitors.

Not only may the proteases ruin the successful protein extraction. If the desired proteins are phosphorylated and the extent and localization of phosphate groups on the protein backbone is the target of the whole analysis, it is necessary to add phosphatase inhibitors to extraction buffer as well. Sodium fluoride inhibits acid phosphatases. Sodium orthovanadate inhibits ATPase, alkaline phosphatase and tyrosine phosphatase. Sodium pyrophosphate and beta-glycerophosphate affect serin/threonin phosphatases.

Recently there also has been an interest in studying ubiquitylated proteins. In this case it is important to use deubiquitylating enzyme inhibitors in the extraction buffer [5].

3.3 Homogenization

The first step of sample preparation is homogenization of multicellular biological material, lysis of defined cell population or clearing fluid samples from debris (cellular or other) and contaminants, such as lipids present in plasma and CSF. Homogeneity or lack thereof may have a profound impact on the final outcome of an entire proteomic experiment; therefore this step should be performed with caution at least equal to all other steps. One of the important sources of analytical variability in homogenization that is difficult to measure is a degree of tissue dispersion. Usually, such procedure is defined by the applied homogenizer working time. For ultrasonication, an introduced power level, homogenization time, and number of cycles is provided. Completeness of bacteria or unicellular organisms' homogenization can be verified by microscopic observations. It is more difficult to measure homogenization of subcellular compartments of eukaryotic cells.

Homogenization is not selective toward sample components and the main goal is to physically disintegrate the sample to release molecular components. Homogenized samples still contain debris, in addition to lipids, saccharides, and metabolites, which all can affect LC and MS separations/signals, thus influencing proteomic profiling. In the subsequent step, samples are usually subjected to centrifugation, yielding a top lipid layer, a middle layer of the soluble proteins and other components and particulate debris sediment at the bottom. This step is usually not validated, besides measurement of protein concentration in the middle layer. Methods of homogenization are summarized in Table 3.1.

It is a good practice to increase reproducibility as well as validation of the results by increased control of the homogenization process by an internal standard addition. Supplementing the sample by adding ("spiking") an internal standard prior to homogenization is a widely accepted practice in analytical chemistry/biochemistry. In the case of samples containing hundreds of proteins, addition of an internal standard, even in the form of another protein, will not ensure that the efficiency of extraction for every compound will be the same. On the other hand, there is no better way to estimate the quantity of extracted proteins from the sample. One caveat in selecting such a perfect internal standard for proteomic experiments is the choice of a protein representing an average characteristic for the entire pool of proteinaceous components. Using bovine serum albumin (BSA) for samples of human origin may provide easily distinguishable spectra and peptides that are detectable at the femtomolar level. On the other hand, albumin is often immunodepleted as it obscures low-abundant proteins, and thus the spiking strategy depends on the particular experimental requirements. It is also important to add the appropriate amount of an internal standard as it is desirable to make precise measurements at low levels.

Table 3.1 The most common methods of biological sample homogenization

Homogenization method	Sample type	Effect
Mechanical, rotor-stator	Most biological tissues, from mild and soft to fibrous, tough. Volumes: Hundreds of microliters to a few liters.	Rough homogenization causing complete tissue disruption. Tissues and cells are usually destroyed. Some subcellular structures might be saved. Used for initial sample homogenization.
Potter-type (PTFE-glass or PTFE—PTFE crushers)	Soft tissues, cell culture material. Volumes: 1—100 ml.	Effect similar to rotor-stator but subcellular organelle is usually destroyed.
Sonication	Very soft tissues, eukaryotic cells from cell culture, microorganisms. Used also for resuspension of pellets and emulsification of nonmixing liquids. Volumes: Single microliters to a few liters.	Very effective if material for the process and ultrasonic power is properly chosen. Possible DNA degradation. Easily overheats sample; effective cooling required.
Liquid nitrogen crushing	Various tissues like tumors, histological biopsy material, animal tissues. Effectiveness depends on the fracturing ability after freezing of the sample. Volumes: No limitations.	Complete degradation of the sample including subcellular structures.
Glass beads shaking/crushing	Microbial and eukaryotic cells in solution, very tiny tissue pieces. Volumes: 1—10 ml.	Effective cell disruption. Subcellular structures remain intact.
Planetary discs blending	Used for fruit tissue or very soft animal tissues, mainly in food processing laboratories. Volumes: 10—1000 ml.	Tissue pulp is formed. Cells are usually not destroyed (allows for cell separation). Addition of enzymes (like trypsin, collagenase) enhances results.
Pressure	Microbial and eukaryotic cells in solution only. Volumes: Continuous process, in general no limitations.	Very effective cell disruption by rapid pressure changes. Possible disruption of subcellular structures.
Lysis buffers	For almost every material. Sample should be prehomogenized (eg, by rotor-stator system) for the best results. Volumes: No limitations.	Effectively destroys cell membranes as well as subcellular membranes of the organelles.

3.4 Homogenization and Isolation of Organelles

Homogenization of the whole tissue will disintegrate cells, in particular if it is performed in a lysis buffer containing detergents, which helps to release proteins. However, many experimental designs seek more information, such as characterization of cells in the tissue. The preferred method of characterizing subsets of cells of the same type is flow cytometry, and therefore a cell suspension would be the material of choice [6]. The advantage of this approach is the possibility of sorting cells to extract a population of interest. One caveat is the low yield of the procedure, which is quite often a major problem with subsequent proteomic profiling. As an alternative, cell cultures may be used in proteomic profiling experiments. Cell culture, which contains only one type of cells, is a more straightforward experimental design, data interpretation and more importantly, experimental validation. Such a system also simplifies analysis; however, it should be taken into consideration that conclusions cannot be simply transferred to the living organisms. Table 3.2 summarizes advantages and limitations of using various types of isolated cell populations.

Cell culturing leads to selection of a desired type of cells based on their different adherence, survival abilities or interactions with toxins or media compounds. Another method, based on physical properties of the cells, is centrifugation in Percoll™ or other polysaccharide-based density gradients, or alternatively, the elutriation technique. Density gradient centrifugation and elutriation are commonly used, especially for separation of blood cell subpopulations. Cell fractions received utilizing centrifugation are significantly enriched in a desired cell subpopulation. The methods are effective, cost-saving, and do not require any sophisticated equipment [7]. The much more cytoselective method of cell separation is also available: cell sorting used in flow cytometry. Technology allows for a very precise separation of cell subpopulations based on, eg, labeling of the cells by antibodies tagged with

Table 3.2 Sources of single cells for proteomic analyses

Source of cells	Comments
Single cell suspension of primary cells	Best reflecting in vivo characteristics of cells. Usually low yield and mixed population.
Cultures of primary cells	Good method to obtain larger number of cells of interest if cells are proliferating in vitro. Cells may change their original phenotype due to culture conditions. If terminally differentiated and nondividing cells are cultured, apoptosis may significantly reduce number of cells, thus will limit amount of material for proteomic analyses.
Cultures of established cell lines	These type of cells are well proliferating, can be obtained in large quantities and constitute a homogenous population. Because such cell lines are transformed they may not correspond to or represent real in vivo situations.
Cells isolated from blood	Can be obtained in relatively large numbers and as 95–98% pure population. Can be further maintained in the in vitro cultures.

fluorescent markers. Instrumentation and preparation of the samples are more expensive than centrifugation in density gradients, but the overall quality of the final sample preparation is much better.

Cells may also be isolated directly from the tissues (like from histopathological or biopsy material) where typical cell culturing or centrifugation methods are unreliable. In such cases, modern microisolation methods are helpful. One of the most advanced techniques is laser capture microdissection, which allows for the isolation of even a single cell from a microtome tissue scrap or any other thin layer of the cells. The operator uses a set of lasers to cut out a single cell or small area of interest from the tissue [8], and the resulting sample is extremely homogenous, but is present only in microscopic quantity, which may limit processing steps for further analysis.

Further steps of (living) cell-containing sample preparation are strongly dependent on the previous purification quality, as well as the selected analytical strategy. However, not every biological material needs additional processing steps after homogenization. For example, when a cell line is homogenized in the presence of a lysis buffer, it can be separated using 2D electrophoresis without additional purification or preparation (contrary to Liquid Chromatography - Mass Spectrometry (LC-MS)). In this case, homogenous cells are easily and completely disrupted by the lysis buffer and typically contain low amounts of lipids, saccharides and DNA that will not interfere with the 2D separation process. Insoluble particles (usually present in high amounts after tissue homogenization) are not observed as the lysis buffer promotes complete degradation and solubilization of cellular and nuclear membranes. Sometimes the appropriately applied homogenization process may be used for isolation of subcellular structures.

As demands have shifted from looking at broad changes in proteomes to investigation of specific metabolic pathways, organellar functions etc., subcellular compartments, such as nuclei, mitochondria, phagosomes, endosomes, etc. are recently being used for proteomic profiling. One caveat is that protocols for organelle extraction allow only for enrichment but not complete purification of the organelle of interest [9]. To obtain subcellular fractions, the homogenate is further processed by differential centrifugation in density gradients. Density gradients, such as sucrose or commercially available kits like Percoll, Histopaque, Ficoll, etc., provide stabilization of the subcellular fractions at the corresponding density of the gradient. Two types of gradients are used:

1. Stepwise gradient, formed by layering the lower-density solution over the higher-density one. This type of gradient is used for separation of the cells (eg, blood cells) based on the differences in their densities, but it may also be used for separation of subcellular structures. Importantly, diffusion can occur at the interface of the gradient zones causing a local disturbance in density and making the final separation less effective. This factor needs to be taken into consideration.

2. Continuous-density gradient, prepared with gradient mixers. In such gradients, upon centrifugation, the subcellular structures are traveling to the region of the same density as their mean, internal densities. In contrast to the stepwise gradients, organelles are not retained in the region dividing two different densities (the density gradient is continuous); however, subcellular fractions of similar densities may penetrate the others, thereby contaminating the purity.

When there is no formation of a visible fraction, so-called "density markers" may be applied. An approach with density markers uses tiny colored beads that localize in a layer of solution density equal to their own or they co-localize with separated organelles (also based on their density). It can help in distinguishing between separated organelles. It should be noted that the gradient-forming agent (eg, sucrose) should be removed from the fraction after centrifugation by either gel filtration or, in case of the cells, washing with a growing medium or balanced salt solution of low density [10]. Finally, the isolated fraction is significantly enriched in one type of subcellular component; however, it often remains contaminated by other structures of similar density (or having the same Svedberg coefficient). Currently, there is no perfect procedure for separation of one specific organelle from the others, but even a somewhat imperfect procedure provides better results when applied as an initial step of the experiment.

3.5 Crude Protein Extraction

There is no single good measure of the quality of crude protein extract. In all cases, protein concentration is measured using various methods based on absorbance at 280 or 220 nm or by colorimetric reactions. These measurements provide a rough estimation of how much protein is present in a given sample; however, they do not imply the quality of the sample. Due to the lack of quality criteria, 1D SDS PAGE is highly recommended in the case of cell lysates, as well as complementary western blot analysis. The latter method raises criticism, as the comparison of the quality of crude samples based on eg, content of actin might not be very accurate. Nevertheless, small molecular metabolites or short peptides may skew protein determinations.

In a vast number of cases, crude samples need to be concentrated with concurrent removal of non-proteinaceous contaminants. Although other methods are available, protein precipitation seems to be the most commonly used [11,12], particularly when the sample is diluted. The preferred methods among precipitants are those using acetone, ethanol, methanol, and their mixtures with trichloroacetic acid (TCA) or sodium deoxycholate. Acetone is the most commonly used solvent, which promotes protein precipitation and simultaneously dissolves nonpolar molecules like lipids. Moreover, it prevents dispersion of proteins among water-based solvents causing protein aggregation and precipitation. Due to the simplicity of acetone application, it is one of the widest used methods for protein precipitation. It has some limitations, such as incomplete precipitation of proteins from diluted samples, but it can be safely used for typical proteomic samples with significant amounts of proteins.

In contrast to acetone application, a mixture of TCA and sodium deoxycholate enhances precipitation of very small amounts of proteins, as deoxycholate binds to the hydrophobic parts of proteins. Addition of TCA increases the hydrophobicity of the deoxycholate–protein complex, which strongly promotes precipitation. However, a combination of TCA with sodium deoxycholate produces sample precipitation that is not suitable for direct MS analysis. Sodium deoxycholate, similar to a majority of detergents, interferes with the ion formation in the ion source of mass spectrometers, and therefore proteins obtained after precipitation must be separated from this detergent. On the other hand, TCA in combination with ethanol (EtOH), is preferentially used if the sample is subsequently separated using SDS-PAGE electrophoresis. This is because the TCA/EtOH mixture, along with protein precipitation, efficiently removes other agents that might be used for sample preparation, such as chaotropic guanidine hydrochloride, which also interferes with ionization in the source.

Technically speaking, protein precipitation is a straightforward procedure, frequently applied for large volumes of samples. The precipitating solvent is added to the sample and then the mixture is left for precipitation at $4°C$ (or even at $-80°C$ for organic solvents, such as ethanol or acetone) for a period ranging from few minutes to few hours. Precipitated proteins are pelleted by centrifugation, washed with cold 70% EtOH, dried in vacuum and resuspended in the solvent suitable for further analyses. The attractiveness of ease and relatively high-throughput protein precipitation is limited by a few drawbacks; the most common are listed next.

1. Certain proteins are more susceptible to precipitation than others; therefore the process may generate quantitative differences between samples in an unpredictable manner.
2. It might be difficult to resuspend the pellet in the typical, water-based solutions unless proteins are resuspended in buffers containing chaotropic agents and/or detergents. This applies to samples destined for separation by 2D SDS-PAGE.
3. Coprecipitation of some contaminants may interfere with downstream analyses.

3.6 Serum and Cerebrospinal Fluid Protein Extraction

Serum/plasma, CSF or other body fluids constitute a separate group of "crude" samples with specific limitations in their analysis. For example, measurement of total protein of plasma/serum or CSF prior to immunodepletion may have in some specific cases little, if any, informative value. However, depletion of the most abundant proteins may also affect the amount of other components, which is why determination of the protein content should be an important step between completion of sample preparation/purification and the beginning of identification of the sample content. Thus, it is recommended to measure protein content at each step of the purification procedure, or add an internal standard before manipulations, or try to find the protein, which may be used as a representative, native standard for the investigated set of samples. In the following sections of this chapter we will review the most commonly used methods for fractionation of crude extracts.

3.7 Fractionation Based on Size-Exclusion Filters

Size-exclusion filters (so-called "cut-off" filters) are membranes with pore sizes specified by the manufacturer, and they function as molecular sieves. They are usually made of cellulose, with low binding of proteins to avoid nonspecific interactions and sample loss. To manufacture these cellulose membranes, nitrocellulose, polyethersulfone (PES) or cellulose triacetate are commonly used. Size-exclusion filters are used in three cases:

1. Removal of salts and other low molecular mass compounds as a faster and easier-to-handle alternative to dialysis or size exclusion chromatography.

2. Concentration of the sample and buffer exchange compatible with the subsequent analytical procedure.

3. Separation of a complex sample into two fractions based on the molecular weight cut-off. This technique should be applied with caution due to the unexpected behavior of proteins. The major question in this case is whether proteins should be denatured partially, completely or not at all prior to their separation by membrane filtration. Mild conditions may promote protein aggregation and nonspecific interactions making complexes of proteins to be split between filtrate and retentate. Therefore, filtration conditions should be carefully optimized for successful separation. An advantage in using this technique is its simple validation using 1D SDS-PAGE.

Fractionation using size-exclusion filters is applied for separating high molecular weight molecules, such as proteins, from low molecular impurities, eg, salts. These filters are also used to concentrate samples when the initial volume is relatively large to be handled by other means, such as vacuum centrifugation or freeze drying. Both these methods (vacuum centrifugation and freeze drying) will lead to concentration of salts, which is often undesirable and also may cause irreversible precipitation of precious material. It should also be noted that fractionation using cut-off should always be considered as a rough method, where often pore size does not correspond to the desired protein molecular mass due to, eg, variabilities in protein shape.

3.8 Chromatographic Methods of Protein Fractionation

Chromatographic techniques are used for desalting, prefractionation, as well as for final separation of the components present in the sample. In the case of desalting, the primary concept of applying chromatographic systems is to separate proteins from other compounds for further analysis. The easiest way of protein separation from nonprotein compounds (salts, DNA, saccharides, lipids) is the application of disposable solid-phase extraction (SPE) microcolumns in a pipette tip. Such columns are able to handle minute quantities of the sample necessary for MALDI-TOF or nano-ESI-MS identification. Larger SPE columns can also be convenient for preparation/purification and preconcentration of proteomic samples. Usage of such types of columns (eg, filled with C2 to C8 RP phase beads) allows for elution of impurities and prefractionation of proteins. Application of a multistep salt gradient on strong cation exchangers (SCX) can fractionate proteins depending on their charge in solution. This simple procedure must ensure compatibility of the eluate with the downstream MS technique. Fortunately, reversed phase (RP) columns can solve the problem of solvent incompatibility in the SCX fractions. Removal of salts on the RP columns is used for fraction purification before MS analysis, thus providing delivery of solution compatible with MS ion sources. Alternatively, salt removal on any column filled with the reversed phase (like C4 or C8) or similar material might be used. It should be noted that any fraction containing salts should not be delivered to the ESI nor MALDI ion sources, as it efficiently disrupts ionization. This also indicates that fractions eluting in the void volume of the RP column should not be directed to the mass spectrometer. Proteins extracted in water-based solvents are hydrophilic, in contrast to proteins present in the cell membrane. Therefore, application of the reversed-phase material for separation may result in the elution of hydrophilic fraction in the void volume or at the very early stage of separation.

Another application of chromatography during proteomic sample preparation is the removal of interfering and unnecessary proteins. These proteins include serum albumin, antibodies, complement C3 and other high abundant proteins in a proteomic sample derived from blood or blood serum. Due to the high abundance, the presence of these proteins can mask those that exist at concentrations that are a few levels of magnitude lower. Thus far, the most effective technique to remove the interfering proteins is application of immunodepletion columns [13]. Immunodepletion methods are described in chapter Immunoaffinity Depletion of Highly Abundant Proteins for Proteomic Sample Preparation.

As an alternative technique for the low-abundance protein enrichment, an "equalizing" method has been reported [14], in which the most abundant proteins are not removed but their concentrations are "equalized" with those at a lower level of concentration. An excess of proteins is removed from the beads. The principle of this approach is based on a library of short peptides bound to a solid support that randomly recognize and bind various proteins, depending on their complementary sequences. As there are equal concentrations of all peptides building the library, every protein but the most abundant can be bound up to the peptide saturation level. The drawback of this approach is that the components cannot be further quantitated (Fig. 3.1).

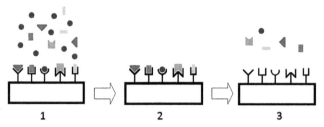

Figure 3.1 Principle of Bio-Rad Proteominer™ methodology. 1. Protein sample containing highly abundant (*blue circles* (dark gray in print versions)) and low abundant (remaining) compounds are mixed on the kit's surface containing peptide map. 2. Proteins from the sample are adsorbed on the peptide map with a quantity not higher than the surface capacity. Excess molecules are removed from the assay. 3. Immobilized proteins are desorbed from the surface. At the moment all proteins are present at a similar concentration.

3.9 Peptide Purification

The methods described so far in this chapter were mainly focused on the extraction and purification of intact proteins. However, depending on the experimental design and goals, identification of all proteins in the sample is not always necessary, and sometimes it is even beneficial to extract a selected set of proteins for subsequent steps. In some cases, analysis of the whole proteome instead of a narrow subset is a disadvantage.

Purification of peptides is usually applied when protein identification using a bottom-up proteomic approach is done with the aid of MS/MS analysis. Due to instrumental limitations, it is still faster and more convenient to sequence and quantitate peptides than proteins. To fulfill mass spectrometer requirements, a mixture of peptides must be free from impurities disrupting MS work. Impurities are usually arising from the previous steps of the analytical procedure. For example, proteins undergoing identification through peptide mapping must be reduced and free cysteine residues must be blocked to avoid accidental disulfide bond formation. Those reactions introduce at least two reagents into the sample (usually dithiothreitol as a reducing agent, and iodoacetamide or 4-vinylpyridine for alkylation). The next step involves introduction of a protease of known activity into the reaction mixture. It is also worth mentioning that each substance requires specific pH, ionic strength, etc. for optimal activity; therefore it is necessary to add buffers, inhibitors or other compounds to the sample. It is obvious that the mixture turns contaminated (fortunately in a controlled manner) and some of the impurities may interact with the ion-formation process during MS-based identification. To solve this problem, the obtained set of peptides is purified *online* during introduction on the chromatographic column in the LC-MS/MS system. Such purification is done automatically by the chromatographic system. Usually an RPC-18 precolumn is used to separate peptides from the contaminants. Under such conditions, the peptides are immobilized on the C18 beads, washed with water-based, acidified solvent, and further directed (eg, switching valve) into the main chromatographic column of the LC-MS/MS system. This approach also prevents introducing salts into MS that are eluted in the void volume.

3.9.1 Phosphopeptides

Another case, being a good illustration where only a fraction of peptides is taken into consideration, is the isolation of a subset that carries specific posttranslational modifications (PTMs), such as phosphorylation [15]. This prompts the question of whether it is more effective to isolate phosphorylated proteins, digest them enzymatically and search for phospho-PTMs, or digest the whole sample, isolate phosphopeptides selectively and identify modified sites. Both strategies can be compared in Fig. 3.2. There are several analytical challenges and neither approach is superior to the other. The strategy strictly depends on the planned goals of the entire experiment.

One of the major challenges is whether we need to obtain and quantitate a ratio between phosphorylated and nonphosphorylated counterparts. If the goal were to maximize identification of phosphorylated peptides, a good approach would be qualitative profiling. In such a case, it is advisable to digest the sample, then pass through, eg, a titanium dioxide

(TiO_2) column, and analyze the enriched phospho-peptide fraction by LC/MS/MS. Though the technical aspect of this approach appears straightforward, the final result is a combination of two sets of samples with their specific analytical characteristics. The TiO_2 flow-through fraction will be much more complex than the bound one, and will likely require an additional step of fractionation, (eg, rotofor fractionation or SCX separation). On the other hand, isolation of all intact phosphoproteins from tissue or cell homogenate needs to be preceded by the sample clean-up to remove impurities that may interfere with TiO_2 chromatography. Between elution of unbound proteins and phosphoproteins, multiple washes of a column with ammonia buffer or other medium is highly recommended. This removes all unbound proteins from the column and the remainder is the pure phosphoproteins fraction. Once a subset of phosphoproteins of desirable purity is achieved, it is subjected to enzymatic digestion, followed by mass spectrometry analysis. It is an analytical challenge to quantify low levels of phosphorylated proteins with the understanding that only 3–5% of the entire pool of a given protein is phosphorylated, and this is all that is needed to induce a biological effect. If the analytical techniques have inherently higher variability than 5%, it will not be possible to measure such change. All of these factors need to be considered before a proteomic experiment is conducted. Another factor is random phosphorylation of proteins, which means that a varying number of phospho-groups might be attached to the particular protein molecule.

3.9.2 Glycopeptides

Besides two major modifications by N- and O-linked glycans, as well as other less common alterations, oligosaccharide chains provide an enormous variability of possible structures. Almost every aspect of this type of modification can undergo variations. The length and the composition of the oligosaccharide chains can be regulated at various places in the protein sequence. A good example of the possible structure complexity is mucins. For

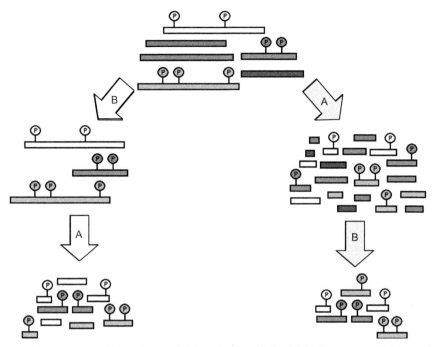

Figure 3.2 Strategies of phosphoprotein identification. Method 1 (*yellow arrows* (light gray in print versions)) uses purification of the phosphoproteins from the sample matrix, followed by enzymatic cleavage of the isolated proteins, and identification of the peptides and phosphopeptides. Method 2 (*blue arrows* (gray in print versions)) applies enzymatic cleavage of all proteins present in the sample with the subsequent isolation of phosphopeptides. Only peptides with phosphate groups are identified. (*Arrows* marked by "A": enzymatic cleavage; *arrows* marked by "B": isolation of the sequences containing phospho-groups.)

example, human mucin-4 (Q99102), present in the human digestive tract, is a protein that is about 2150 amino acids long. According to UniProtKB database, it has at least 23 known glycosylation sites in its structure. Depending on the organism requests, mucin can be partially or fully glycosylated with different oligosaccharide chains. This leads to a very high level of variability between function, localization, and activity of the protein molecule and, of course highly complicates analysis of such proteins.

There are many approaches for sample preparation to analyze glycosylation sites; however the use of lectin affinity columns is a method of choice, as specific lectins on a solid support can pull out proteins carrying modifications. Besides obstacles resulting from a broad intricacy of the sample, lectin affinity chromatography used as a method of choice is very efficient in purification of glycopeptides from the sample matrix. This analytical approach supports identification of glycopeptides and can be used as a subsequent validation of the applied methodology.

3.10 Detergents, Lipids and DNA

While working with biological material, three major groups of substances are of importance and should not be ignored during sample preparation: detergents, lipids, and nucleic acids. Detergents are usually applied for extraction of proteins, in particular those originating from cellular membranes or to improve solubilization of proteins. Due to their chemical properties, detergents can unfavorably influence the whole analytical workflow. The remaining substances, such as lipids and nucleic acids, should also be eliminated from protein samples because they may contribute to a series of analytical problems.

3.10.1 Detergents

Detergents are amphiphilic molecules, as they possess hydrophobic and hydrophilic parts in their structures. The hydrophobic group usually consists of a hydrocarbon tail while the hydrophilic part has a polar head. In a water environment, detergents, if they have been added at the appropriate concentrations, are capable of forming micelles. The concentration allowing for micelle formation in water solution is called CMC (critical micelle concentration) and is an important factor during protein solubilization. Too low a concentration of the detergent would result in a poor recovery of membrane proteins. On the other hand, too high a concentration may impact the quality of final results, as the removal of the detergent, present in excess, is difficult and may lead to unpredictable protein loss from the sample. Detergent concentration close to the CMC can be determined empirically by physical measurements, eg, surface tension, or by chemical methods. The most convenient technique is addition of dyes changing the color while micelles are formed.

Detergents can be divided into three main groups: nonionic, ionic and zwitterionic. Nonionic detergents, dissolved in water, do not have charge in the area of the hydrophilic head. They are especially effective in breaking interactions between lipids or lipids and proteins, in contrast to their inability to break protein–protein interactions. Typical representatives of this group are tritons (X-100, X-114), BRIJs, SPANs, MEGAs, NP-40 or tweens. Ionic detergents possess a stable charge located on the hydrophilic part, when dissolved in water solutions. A typical representative of this group is sodium dodecyl sulfate (SDS), one of the most widely used detergents among laboratories due to its properties. Similarly to other ionic detergents, SDS binding to proteins supplements them with multiple negative charges, which mask native charges of the amino acid residues. This is the reason for the routine use of SDS-PAGE electrophoresis. Other known detergents in this group are deoxycholic acid, sarcosyl (sodium lauroyl sarcosinate), and others. The last group, zwitterionic detergents, are the substances simultaneously possessing positive and negative charges but their net charge is equal to zero. Zwitterionic detergents do not change their own charge while solubilizing proteins. They are also more useful for solubilization of the proteins than nonionic detergents, mainly due to the inhibition of protein–protein aggregation. According to their properties, zwitterionic detergents are used during isoelectrofocusing or in 2D electrophoresis. Typical representatives of this group are: CHAPS, CHAPSO and sulfobetaines.

Another important factor of detergents is the so-called "cloud point" (CP). This phenomenon depends on the temperature of the medium in which detergent is dissolved. CP is mainly observed for nonionic detergents, but others (eg, SDS, CHAPSO) have also their own cloud points at higher temperatures, usually close to 90–100°C (purification of membrane or cytosolic proteins). The name of the phenomenon is taken from the "clouds" formed in the detergent solution after reaching its cloud point temperature (CPT). The solution separates into two phases: (1) the aqueous part, mainly containing water-dissolved proteins; and (2) the "cloudy" part containing detergent with hydrophobic proteins. Such a nonhomogenous mixture can be easily separated, which allows for an easy partition of the solubilized proteins into hydrophilic and hydrophobic ones. Some detergents have relatively low cloud point temperatures, like Triton X-114 (23°C), which favors them to be applied in routine laboratory applications. It should be clearly stated at this point that detergents may significantly affect proteolytic digestion and MS measurements. Therefore, removal of these agents is often obligatory before further analysis.

It is important to mention that detergents can work not only in the water solutions but also in the organic solvents, where they also can form micelles, but the polar "heads" of the detergent molecules are located inside the micelle, while the hydrophobic "tails" are directed toward the solvent, thus allowing for a broader applicability of detergents in various experiments, including proteomics.

Detergents can cause further analytical problems due to their physicochemical properties. They are usually prone to foaming when mixed or stirred. As mentioned above, they can also lead to mass spectra disruption or sensitivity loss, or they may influence HPLC separations by nonspecific interactions with the column stationary phase. Detergent removal seems to be a difficult task, leading to the protein losses during the process. To remove detergents, precipitation with the aid of acetone or similar solvents is used, but for their low content, the reversed phase (eg, disposable cartridges) is frequently applied. Nowadays, there are also highly efficient detergent removal kits available on the market.

3.10.2 Lipids and DNA

Lipids and DNA, being typical compounds of the animal tissues, can interact with peptides/proteins present in the sample, which may lead to low-quality results. Interestingly, RNA is more susceptible to endogenous enzymes (RNAses) and is rapidly degraded during initial preparation steps. Small amounts of lipids, eg, in samples obtained from the cell culture, usually do not have a significant impact on the data acquisition and final results quality. Unluckily, during tissue processing (eg, brain, spinal cord), significant amounts of lipids may cause problems with 1D or 2D electrophoretic separation, HPLC, etc. There are procedures described in the literature on how to remove DNA and lipids from the proteomic samples. The most common procedure of lipid removal is precipitation of proteins from the organic solvent solubilizing lipids. In this case, sample containing proteins and lipids is mixed with the organic solvent, allowing for precipitation of the proteins without coprecipitation of the lipids. After the sample has been centrifuged, the pellet is dried, washed to remove residual impurities, and dissolved in a water-based solvent. The following solvents are usually applied for precipitation: acetone, methanol, and trichloroacetic acid alone or with a coprecipitant like deoxycholic acid sodium salt.

DNA causes analogous difficulties during proteomic sample processing. It may interact with chromatographic, and especially electrophoretic separations. DNA, similarly to lipids, can evoke bars (1D electrophoresis) or spots (2D electrophoresis) relocation and streaking. Therefore, if it is present in the sample in significant amounts, it should be removed at the early stage of sample preparation. DNA can be removed during the protein precipitation procedure along with lipids. In some cases DNA removal is only partial, and sometimes the removed quantity is insufficient for high-quality electrophoretic separation. One of the most efficient DNA precipitation methods was described by Antonioli et al. [16], where a mixture of phenol, chloroform, and isoamyl alcohol were applied to solubilize DNA and precipitate the protein fraction. There is also another technique to remove DNA from the sample, but it needs addition of the external protein into the mixture. Deoxyribonucleases are a class of endonucleases, selective toward DNA. If added to the cell or tissue homogenate, they can efficiently destroy DNA by truncating it into short sequences. Addition of a small amount of such an endonuclease at the early stage of sample preparation can solve the problem associated with DNA. However, such addition of the external protein always results in sample contamination and, when added in excess, may also obscure identification of other endogenous components. DNAse identification in the sample should be excluded from the identified proteins list. One of the commercial preparations used for this purpose is an enzyme distributed under the name Benzonase™ (Merck Millipore), which is a recombinant endonuclease from *E. coli*.

3.11 Summary

All steps of sample preparation and fractionation for proteomic studies are an integral part of quantitative or qualitative profiling. The strategy of prefractionation depends on the desired results, as well as on the sample type. Although the subsequent steps are not stringently connected to the preceding steps, the quality of the sample has a critical impact on the final outcome. Reproducibility of extraction after tissue/cell/organelle disintegration, yield and integrity of proteins, are factors that need to be constantly monitored and attention needs to be paid to the diligence of operators in maintaining parameters and executing all the steps. This is even more important when hybrid techniques of separation are applied favoring isolation and/or purification of specific subsets of proteins. As knowledge expands, the issues addressing the biological importance of proteomic profiling will unquestionably have an impact on every step of a proteomic experiment.

Acknowledgments

The authors would like to thank the Polish National Science Center, as this work was partially supported by the grants of the NSC (grant numbers 2012/07/B/NZ4/01468 and 3048/B/H03/2009/37) and EuroNanoMed "META", grant number: 05/EuroNanoMed/2012.

References

[1] Tang J. Acid proteases come of age. Nature 1977;266:119−20.

[2] Beynon R, Bond JS. Proteolytic enzymes: a practical approach. Oxford, United Kingdom: IRL Press at Oxford University Press; 1989.

[3] Fahrney DE, Gold AM. Sulfonyl fluorides as inhibitors of esterases. 1. Rates of reaction with acetylcholinesterase, aplha-chymotripsin, and trypsin. J Am Chem Soc 1963;85:997−1000.

[4] Salama ZB. Studies on the influence of various effectors on proteinases of rat-liver lysosomes in vitro. Acta Biol Med Ger 1980;39:355−66.

[5] Altun M, Kramer HB, Willems LI, McDermott JL, Leach CA, Goldenberg SJ, et al. Activity-based chemical proteomics accelerates inhibitor development for deubiquitylating enzymes. Chem Biol 2011;18:1401−12.

[6] Shapiro HM. Practical flow cytometry. J. Wiley & Sons; 2003.

[7] Tienthai P, Kjellen L, Pertoft H, Suzuki K, Rodriguez-Martinez H. Localization and quantitation of hyaluronan and sulfated glycosaminoglycans in the tissues and intraluminal fluid of the pig oviduct. Reprod Fertil Dev 2000;12(3−4):173−82.

[8] Murray GL, Curran S. Laser capture microdissection. Humana Press; 2005.

[9] Kuster DW, Merkus D, Jorna HJ, Dekkers DH, Duncker DJ, Verhoeven AJ. Nuclear protein extraction from frozen porcine myocardium. J Physiol Biochem June 2011;67(2):165−73.

[10] Li X, Donowitz M. Fractionation of subcellular membrane vesicles of epithelial and nonepithelial cells by OptiPrep density gradient ultracentrifugation. Methods Mol Biol 2008;440:97−110.

[11] Evans DR, Romero JK, Westoby M. Concentration of proteins and removal of solutes. Methods Enzymol 2009;463:97−120.

[12] Burgess RR. Protein precipitation techniques. Methods Enzymol 2009;463:331−42.

[13] Moser AC, Hage DS. Immunoaffinity chromatography: an introduction to applications and recent developments. Bioanalysis April 2010;2(4):769−90.

[14] Boschetti E, Righetti PG. The ProteoMiner in the proteomic arena: a non-depleting tool for discovering low-abundance species. J Proteomics August 21, 2008;71(3):255−64.

[15] Eyrich B, Sickmann A, Zahedi RP. Catch me if you can: mass spectrometry-based phosphoproteomics and quantification strategies. Proteomics February 2011;11(4):554.

[16] Antonioli P, Bachi A, Fasoli E, Righetti PG. Efficient removal of DNA from proteomic samples prior to two-dimensional map analysis. Journal of Chromatography A 2009;1216:3606−12.

4

PROTEIN EXTRACTION AND PRECIPITATION

P. Novák and V. Havlíček
Institute of Microbiology, Academy of Sciences of the Czech Republic, Prague, Czech Republic

CHAPTER OUTLINE
4.1 Introduction   51
4.2 Focus on Hydrophobic Protein Extraction   52
4.3 The Role of Protein Solvation   53
4.4 Protein Precipitation   55
4.5 Salting Out   55
4.6 Isoelectric Point Precipitation   56
4.7 Organic Solvent-Driven Precipitation   57
4.8 Trichloroacetic Acid Precipitation   60
Acknowledgment   61
References   61

4.1 Introduction

Efficient extraction of proteins is a key factor for successful proteomic experiments. Depending on the source, proteins have to be brought into solution by breaking the tissue or cells or extracted from various sources. To extract proteins from solid samples it is necessary to disintegrate the samples first. There are several ways to perform this task: repeated freezing and thawing, sonication, homogenization by high pressure, filtration, permeabilization by organic solvents or hypotonic shock. There are several methods to achieve this. In the case of tissues and cells, each method consists of three steps: sample disintegration, protein extraction and precipitation. It is very important to select the proper workflow

Proteomic Profiling and Analytical Chemistry. http://dx.doi.org/10.1016/B978-0-444-63688-1.00004-5

right from the beginning. All proteomic experiments use mass spectrometry as a final analytical tool for peptide sequencing and protein identification. The best sample is represented by a pure protein in a solid state or volatile buffer without any additives such as salts, surfactants (mainly detergents) and matrix components coming from the biological sample (lipids, metabolites, nucleic acids and oligosaccharides). In reality, preparation of such a sample is difficult if possible at all. Although small amounts of contaminants can be tolerated in such analyses, some proteins are unstable and tend to precipitate out from solution if not kept in a proper environment represented by some concentration level of a detergent, chaotropic agent and/or salt. There are two main considerations consider when designing proteomics experiments:

1. Which types of proteins (membrane proteins, nuclear proteins, cytosolic proteins or secreted proteins, glycoproteins) are being analyzed.
2. Which analytical technique (centrifugation, liquid chromatography or electrophoresis) represents a prerequisite fractionation step prior to mass spectrometric analysis. The method of choice depends on protein fragility and the sturdiness of the cells. Soluble proteins will remain in the solvent after the extraction process and can be separated from cell membranes and nucleic acids by centrifugation. Taking this all together, it implies the utilization of sample preparation protocol for direct analysis of hydrophobic membrane proteins and a completely different procedure for soluble secreted proteins.

4.2 Focus on Hydrophobic Protein Extraction

Phenol and chloroform/methanol extractions represent alternatives suitable particularly for hydrophobic proteins. Phenol dissolves proteins and lipids, leaving water-soluble matter (carbohydrates, nucleic acids, etc.) in the aqueous layer [1]. Particulate and "ambiguous" matter remains insoluble. Phenol extraction of nucleoproteins (eg, virus particles) gives pure products. Crude tissue yields complex mixtures, particularly in the aqueous phase. One of the major challenges in proteomics concerns membrane proteins. A procedure based on the differential extraction of membrane proteins in chloroform/methanol mixtures was tested on two different chloroplast membrane systems: envelope and thylakoid membranes. The propensity of hydrophobic proteins to partition in chloroform/methanol mixtures directly correlates with the number of amino acid residues (Res) per number of putative transmembrane (TM) regions (Res/TM ratio). Regardless of special cases of some lipid-interacting proteins, chloroform/methanol extractions allow for enrichment of hydrophobic proteins and exclusion of hydrophilic proteins from both membrane systems [2].

4.3 The Role of Protein Solvation

In most instances, extracted proteins need to be concentrated prior to subsequent analytical steps of a proteomic experiment. There are several techniques available, such as ultrafiltration, affinity chromatography and/or precipitation. In contrast to protein precipitation, the other techniques exhibit relatively poor recovery. Despite the development of new materials used in ultrafiltration, a significant amount of proteins can nonspecifically stick to ultracentrifugation membranes. Proteins may also stick to some degree to stationary phases used for affinity chromatography. Precipitation is believed to yield the highest recovery and it is suitable for low-volume protein extracts. The basic principle of precipitation is to alter the solvation potential of the solvent and thus lower the solubility of the solute by addition of a reagent. The solubility of proteins in aqueous buffers depends on the distribution of hydrophilic and hydrophobic amino acid residues on the protein surface. Hydrophobic residues predominantly occur in the globular protein core, but some exist in patches on the surface. Proteins having high hydrophobic amino acid content on the surface have low solubility in an aqueous solvent. Charged and polar surface residues interact with ionic groups in the solvent and increase solubility. Knowledge of amino acid composition of a protein leads to determination of an ideal precipitation solvent and method. Repulsive electrostatic forces are in place during dissolving proteins in an electrolyte solution. These repulsive forces among proteins prevent aggregation and facilitate dissolution. Solvent counter ions migrate toward charged surface residues on a protein, forming a rigid matrix of counter ions attached to the protein surface. The adjacent solvation layer, which is less rigid, consists of a decreasing concentration profile of the counter ions and an increasing concentration profile of the co-ions. In effect, the potential of proteins to engage in ionic interactions with each other will decrease, resulting in lower chances of forming aggregates (Fig. 4.1).

Water molecules can have a similar effect, as they forms a solvation layer around hydrophilic surface residues of a protein, establishing a concentration gradient around the protein, with the highest concentration at the protein surface. This water

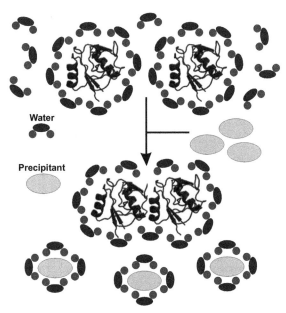

Figure 4.1 The basic principle of the protein precipitation mechanism.

network has a damping effect to the attractive forces between proteins. Dispersive or attractive forces exist between proteins through permanent and induced dipoles. For example, basic residues on a protein can have electrostatic interactions with acidic residues on a different protein molecule. However, solvation by ions in an electrolytic solution or water will decrease protein–protein attractive forces. Decreasing the hydration layer by adding reagents around the protein enhances protein accumulation and precipitation.

4.4 Protein Precipitation

Protein precipitation occurs in a stepwise manner. The addition of a precipitating agent and steady mixing, causing the precipitant and target to collide, destabilizes the protein solution. Sufficient mixing time is required for molecules to diffuse across the fluid eddies. During the following nucleation phase, submicroscopic-sized particles are generated and the growth of these particles is under Brownian diffusion control. Once the growing particles reach a critical size (0.1–10 μm for high and low shear fields, respectively), by diffusive addition of individual protein molecules, they continue to grow by colliding into each other and sticking or flocculating. This phase occurs at a slower rate than mixing the precipitant. During the final step, aging in a shear field, the precipitate particles repeatedly collide and stick, then break apart, until reaching a stable mean particle size, which is dependent upon individual proteins. The mechanical strength of the protein particles correlates with the product of the mean shear rate and the aging time. Aging helps the particles to withstand the fluid shear forces encountered in pumps and centrifuge feed zones without reducing in size [3].

4.5 Salting Out

Salting out is a spontaneous process when the appropriate concentration of the salt is reached in solution. The hydrophobic patches on the protein surface generate highly ordered water shells. The ideal salt for protein precipitation is most effective for a particular amino acid composition, is inexpensive, and is nonbuffering. The most commonly used salt is ammonium sulfate. The common experimental set up consists of two steps. In the first step, the nucleic acids are precipitated with low concentration of ammonium sulfate (around 30%, w/v), the precipitate is spun down and proteins are salted out using 65% w/v ammonium sulfate. There is a low variation in salting-out temperatures ranging between 0°C and 30°C.

Salting out, though rarely used in proteomic studies, provides refolded/native protein conformation after salt removal and significantly reduces sample volume. This approach is useful if, eg, active enzymes are the subject of profiling. Addition of a neutral salt compresses the solvation layer around proteins and increases protein–protein interactions. As the salt concentration of a solution increases, more of the bulk water becomes associated with the salt ions. As a result, fewer water molecules are available to the solvation layer around the protein molecule, which exposes hydrophobic patches to the protein surface. Proteins may then exhibit hydrophobic interactions, aggregate and precipitate from solution. Protein precipitates left in the salt solution can remain stable for years, protected from proteolysis and bacterial contamination by the high salt concentrations. This precipitation procedure purifies the proteins in their native form while maintaining the activity of a protein, particularly in the case of enzymes.

Polymers, such as dextrans and polyethylene glycols, can also precipitate proteins, and they are less likely to denature the biomaterials than isoelectric precipitation. These polymers in solution attract water molecules away from the solvation layer around the protein increasing the protein–protein interactions and enhancing precipitation.

4.6 Isoelectric Point Precipitation

The isoelectric point (pI) is the pH of a solution at which the net charge of a protein becomes zero. At solution pH that is above the pI, the surface of the protein is predominantly negatively charged, and therefore like-charged molecules will exhibit repulsive forces. Likewise, at a solution pH that is below the pI, the surface of the protein is predominantly positively charged, and repulsion between proteins occurs. However, at the pI, the negative and positive charges are balanced, reducing repulsive electrostatic forces, and the attraction forces predominate, causing aggregation and precipitation. The pI of most proteins is in the pH range of 4 to 7. Mineral acids, such as hydrochloric and sulfuric acids, are used as precipitants. The greatest disadvantage of isoelectric point precipitation is the irreversible denaturation caused by the mineral acids. For this reason isoelectric point precipitation is most often used to precipitate contaminant proteins rather than the target protein [4].

4.7 Organic Solvent-Driven Precipitation

Addition of miscible organic solvents such as ethanol, methanol or acetone to a solution may cause proteins in solution to precipitate. The solvation layer around the protein will decrease as the organic solvent progressively displaces water from the protein surface and binds it in hydration layers around the organic solvent molecules. With smaller hydration layers, the proteins can aggregate by attractive electrostatic forces. Important parameters to consider are temperature, which should be less than $0°C$ to avoid denaturation, pH and protein concentration in solution. Miscible organic solvents decrease the dielectric constant of water, which in effect allows two proteins to come close together. There are several examples of using ethanol precipitation in proteomic experiments. There is an interesting study showing ethanol precipitation of human plasma [5] where lipids and very-low-density lipoproteins were pelleted by centrifugation for 15 min at $15,000 \times g$; IgG were removed using protein G Sepharose beads; and finally serum albumin was precipitated using 42% ethanol at $4°C$ for 1 h. Currently such a method is utilized to provide proteomic experiments on body fluids like serum and follicular and cerebrospinal liquids. It is known that these body fluids contain a high amount of so-called "ballast proteins" (almost 95%), eg, serum albumin, immunoglobulins, transferrin, etc., and scientific interest focuses on the remaining 5% of proteins containing potential markers. Immunodepletion of those highly abundant proteins is time-consuming and expensive (for more details, see Chapter 6). The alternative to fractionation is ethanol precipitation, and Elssner et al. reported that 60–80% v/v ethanol concentration predominantly precipitates the ballast proteins, leaving the remaining "interesting" protein fraction in solution [6].

Another use of ethanol/methanol precipitation is glycoprotein analysis. So often N-linked oligosaccharides are removed enzymatically/chemically from the protein backbone and deglycosylated protein is precipitated by 70% ethanol/methanol. Released oligosaccharides remain in solution for subsequent mass spectrometric analysis. Also it is very important to keep in mind that alcohols (methanol, ethanol, isopropanol) precipitate proteins. For example, the mixture of acidified aqueous solution of methanol is very popular within mass spectrometry researchers. Peptides and some proteins ionize well under these conditions. On the other hand, most proteins precipitate and clog the spraying needle. The alternative to alcohol precipitation is the application of acetone, providing even better protein recovery. Usually the sample is mixed with ice-cold acetone to 70% v/v final concentration and kept at $-20°C$ for at least 1 h. In contrast to alcohol precipitation, the procedure is not selective, and it more or less precipitates all proteins in the sample.

The current proteomics effort is mainly driven by biomarker discovery, and human body fluids represent the most-investigated specimens. As mentioned earlier, the alcohols may precipitate ballast proteins, keeping the biologically interesting molecules in solution for further analysis. Since peptides and lower-sized proteins (approximately below 40 kDa, eg, cytokines and growth factors) are the most biologically significant biomarkers, the alternative approaches were described for their enrichment from human body fluids, mainly human sera and plasma, including precipitation of midsize and large protein molecules by acetonitrile. Acetonitrile is miscible with water in all ratios; therefore small proteins and peptides are soluble in acetonitrile-containing buffers. Chertov et al. described rapid dilution of mouse sera with two volumes of acetonitrile containing 0.1% TFA as an ion-pairing agent to disrupt peptides and smaller proteins from large and abundant proteins, thereby facilitating extractions [7]. This approach was further transferred from mouse model to human sera with minor changes. Later it was demonstrated that the addition of TFA is not necessary for small protein biomarker discovery [8]. The recent modification of this method represented the addition of ammonium bicarbonate buffer. The supernatant containing low abundant and small proteins was further delipidated by liquid–liquid extraction to increase the sensitivity of subsequent mass spectrometric analysis [9]. Solid-phase extraction is a nonselective depletion method of midsize and large proteins. In one strategy, human plasma was diluted 10 times in 50-mM ammonium bicarbonate at neutral pH, acidic (0.1% formic acid) or basic (0.3% ammonia) conditions. These samples were incubated with porous solid stationary phase, where peptides and small proteins were trapped to particle surface via hydrophobic interaction. The unbound proteins are washed out and molecules of interest were later eluted with acidic 75% acetonitrile (0.1% formic acid). The basic conditions worked best for small protein enrichment [10]. A second strategy was very similar to size-exclusion solid-phase surface enrichment, but instead of a hydrophobic surface, the peptides and small proteins entering the solid particles were trapped to porous core–shell hydrogel nanoparticles as high-affinity baits. Human plasma and serum samples were diluted 1:2 with 50 mM Tris-Cl buffer (pH 7). Peptides and small proteins entered the core–shell hydrogel, in which they were trapped by high-affinity binder. The particles with bound proteins were centrifuged and the captured molecules were eluted with appropriate buffer regardless of the type of core–shell particles [11]. The third extraction approach utilized the hydrogel as well, and was used for low abundant protein pre-concentration. From the low concentrated plasma, hydrogel particles captured abundant proteins at higher efficiency than low abundant ones, which were enriched in the supernatant, whereas hydrogel particles incubated with high concentrations of plasma captured and irreversibly trapped abundant proteins. During elution, the irreversibly trapped proteins remained captured on hydrogel, while low abundance proteins were released and recovered in the eluate.

For example, hydrogel particles based on poly(N-isopropylacrylamide) (pNIPAm) are well-characterized water-swellable polymers and belong to family of anionic pH-sensitive hydrogels containing carboxylic groups such as acrylic acid (pNIPAm-AAc). This process leads to hydrogel shrinking and concomitant reduction of particle diameter and pore size. As the cationic proteins bind the hydrogel anionic acid groups, the particles experience a continuous shrinking equivalent to the addition of an acid [12].

4.8 Trichloroacetic Acid Precipitation

Trichloroacetic acid (TCA) is a very effective protein-precipitating agent, especially for precipitating proteins from dilute solutions. Surprisingly, little is known about the precipitation mechanism. However, it was shown that the relationship between precipitation extent and TCA concentration can be described as a U-shaped curve with optimum precipitation at roughly 15% TCA, suggesting hydrophobic aggregation [13] as the dominating mechanism [14]. Bensadoun and Weinstein reported that a different carrier molecule, deoxycholate (DOC), working at neutral pH, also improves protein recovery [15]. Instead of DOC, any non-ionogenic detergent (Nonidet 40, Triton X100, etc.) may also be used. In general, the detergent is added to a protein sample to reach 1% w/v concentration. Finally, 30% aqueous solution of TCA is introduced to a protein sample to achieve 10−15% w/v concentration. The precipitation occurs on ice for few hours. It is important to note the protein precipitate is extremely acidic and contains surfactants. In all cases it is necessary to extract remaining TCA and the detergent, as they are both detrimental to mass spectrometric analysis. TCA denatures proteases in a shotgun experiment and affects the electrophoretic separation. Thus it is recommended to wash the protein pellet several times with ice-cold acetone or an ice-cold mixture of ethanol/ethyl ether following TCA precipitation and prior to further proteomic protocols.

Acknowledgment

The authors acknowledge the support from the Ministry of Education, Youth and Sports of the Czech Republic (NPU LO1509).

References

[1] Cohn EJ, C JB. The molecular weights of proteins in phenol. Proc Natl Acad Sci 1926;12:433−8.

[2] Ferro M, Seigneurin-Berny D, Rolland N, Chapel A, Salvi D, Garin J, et al. Organic solvent extraction as a versatile procedure to identify hydrophobic chloroplast membrane proteins. Electrophoresis 2000;21:3517−26.

[3] http://en.wikipedia.org.

[4] Zellner M, Winkler W, Hayden H, Diestinger M, Eliasen M, Gesslbauer B, et al. Quantitative validation of different protein precipitation methods in proteome analysis of blood platelets. Electrophoresis 2005;26:2481−9.

[5] Fu Q, Garnham CP, Elliott ST, Bovenkamp DE, Van Eyk JE. A robust, streamlined, and reproducible method for proteomic analysis of serum by delipidation, albumin and IgG depletion, and two-dimensional gel electrophoresis. Proteomics 2005;5:2656−64.

[6] Elssner T, Pusch W, Shi G, Kostrzewa M. High-molecular weight protein depletion strategy for enhanced MALDI-TOF MS serum peptide profiling. Mol Cell Proteom 2006;5:S265.

[7] Chertov O, Biragyn A, Kwak LW, Simpson JT, Boronina T, Hoang VM, et al. Organic solvent extraction of proteins and peptides from serum as an effective sample preparation for detection and identification of biomarkers by mass spectrometry. Proteomics 2004;4:1195−203.

[8] Kay RG, Barton C, Velloso CP, Brown PR, Bartlett C, Blazevich AJ, et al. High-throughput ultra-high-performance liquid chromatography/tandem mass spectrometry

quantitation of insulin-like growth factor-I and leucine-rich alpha-2-glycoprotein in serum as biomarkers of recombinant human growth hormone administration. Rapid Commun Mass Spectrom 2009;23:3173−82.

[9] Such-Sanmartin G, Bache N, Callesen AK, Rogowska-Wrzesinska A, Jensen ON. Targeted mass spectrometry analysis of the proteins IGF1, IGF2, IBP2, IBP3 and A2GL by blood protein precipitation. J Proteom 2015;113:29−37.

[10] Barton C, Kay RG, Gentzer W, Vitzthum F, Pleasance S. Development of high-throughput chemical extraction techniques and quantitative HPLC-MS/MS (SRM) assays for clinically relevant plasma proteins. J Proteome Res 2010;9:333−40.

[11] Tamburro D, Fredolini C, Espina V, Douglas TA, Ranganathan A, Ilag L, et al. Multifunctional core-shell nanoparticles: discovery of previously invisible biomarkers. J Am Chem Soc 2011;133:19178−88.

[12] Such-Sanmartin G, Ventura-Espejo E, Jensen ON. Depletion of abundant plasma proteins by poly(N-isopropylacrylamide-acrylic acid) hydrogel particles. Anal Chem 2014;86:1543−50.

[13] Sivaraman T, Kumar TKS, Jayaraman G, Yu C. The mechanism of 2,2,2-trichloroacetic acid-induced protein precipitation. J Protein Chem 1997;16:291−7.

[14] Xu Z, Xie Q, Zhou HM. Trichloroacetic acid-induced molten globule state of aminoacylase from pig kidney. J Protein Chem 2003;22:669−75.

[15] Bensadoun A, Weinstein D. Assay of proteins in presence of interfering materials. Anal Biochem 1976;70:241−50.

5

ONLINE AND OFFLINE SAMPLE FRACTIONATION

M. Smoluch, P. Mielczarek and A. Drabik
AGH University of Science and Technology, Krakow, Poland

J. Silberring
AGH University of Science and Technology, Krakow, Poland;
Polish Academy of Sciences, Zabrze, Poland

CHAPTER OUTLINE

5.1 INTRODUCTION 65

5.2 STRONG CATION EXCHANGE, WEAK CATION EXCHANGE, CONTINUOUS OR STEP GRADIENT? 66
5.2.1 Historical Perspective 66
5.2.2 Principle of Ion Exchange Chromatography 66
5.2.3 Common Types of Ion Exchange Chromatography Stationary Phases 68
5.2.4 Choice of Ion Exchanger (Cation or Anion?) 71
5.2.5 Choice of Strong or Weak Ion Exchanger 73
5.2.6 Buffers in Ion Exchange Chromatography 73
5.2.7 Ion Exchange Chromatography in Proteomic Studies 74
References 76

5.3 PROTEIN AND PEPTIDE SEPARATION BASED ON ISOELECTRIC POINT 77
5.3.1 Principles of Isoelectric Focusing 77
5.3.2 Sample Preparation Prior to Isoelectric Focusing 80
5.3.3 Isoelectric Focusing in Liquid State 82
5.3.4 Immobilized pH Gradient Isoelectric Focusing 83
5.3.5 Capillary Isoelectric Focusing 83
5.3.6 Isoelectric Focusing in Living Organisms 84

Proteomic Profiling and Analytical Chemistry. http://dx.doi.org/10.1016/B978-0-444-63688-1.00005-7
63

5.3.7 Summary   85
References   85

 **5.4 CAPILLARY COLUMNS FOR PROTEOMIC
    ANALYSES   86**
5.4.1 Introduction   86
5.4.2 Conventional Capillary Columns   87
5.4.3 Monoliths   89
  5.4.3.1 Silica-Based Monoliths   89
  5.4.3.2 Organic-Based Monoliths   91
  5.4.3.3 Methacrylate-Based Monoliths   92
  5.4.3.4 Styrene-Based Monoliths   92
5.4.4 Summary and Conclusions   94
5.4.5 Recent Developments   96
References   98

INTRODUCTION

M. Smoluch
AGH University of Science and Technology, Krakow, Poland

Fractionation is usually performed to obtain the best possible protein profile from very complex samples. The output is usually expressed as the number of proteins in the resulting analysis. Two strategies of fractionation can be applied at the protein or peptide level. The selection of the strategy is not an easy task because it depends on many factors. An *online* approach can give comparable results to the *offline* strategy, but without extra sample-handling time. Each additional step in the offline sample fractionation increases the risk of sample loss, and that may lead to a lower protein coverage. An online fractionation is very often a compromise between methods being coupled. This compromise means that the methods coupled online do not operate at maximal possible performance. Several strategies of sample fractionation are subjects of this chapter.

5.2

STRONG CATION EXCHANGE, WEAK CATION EXCHANGE, CONTINUOUS OR STEP GRADIENT?

P. Mielczarek
AGH University of Science and Technology, Krakow, Poland

J. Silberring
*AGH University of Science and Technology, Krakow, Poland;
Polish Academy of Sciences, Zabrze, Poland*

5.2.1 Historical Perspective

Historically, stationary phases for ion exchange chromatography (IEC) were available as large beads and used predominantly for purification of crude products at low pressure, and at flow rates ranging from 1.0 to 2.0 mL/min. A breakthrough in IEC started in 1951, when this technique was used for the first time to separate mixtures of amino acids [1]. The analysis was performed under step gradient conditions and specific temperature program that lasted roughly 600−700 min. Nevertheless, the need for this type of analysis led to a rapid design of specialized instrumentation termed "amino acid analyzers." At the present time, IEC is mostly used in the high-performance/high-pressure liquid chromatography (HPLC) format.

5.2.2 Principle of Ion Exchange Chromatography

IEC is a type of chromatography where ions or polar molecules can be separated by their interactions (mostly by reversible adsorption) with oppositely charged ion exchange groups immobilized on an insoluble support. The mobile phase in IEC is aqueous because the formation of ions is favored in such solutions and buffers are usually adjusted to a particular pH. The mechanism of an ion exchange process for both positively and negatively charged ions can be represented by the following equations:

$$\left(R^- - Y^+\right)_s + \left(X^+\right)_m \rightarrow \left(R^- - X^+\right)_s + \left(Y^+\right)_m$$

$$\left(R^+ - Y^-\right)_s + \left(X^-\right)_m \rightarrow \left(R^+ - X^-\right)_s + \left(Y^-\right)_m$$

where s indicates state for stationary phase and m indicates mobile phase. One ion can be separated from the other in the mixture providing the stationary phase has selective affinity for the various analyte ions.

Most separations based on IEC involve five steps to complete the procedure presented in Fig. 5.2.1.

Always there is a risk that, besides an ion exchange effect, some nonspecific adsorption of the analyte(s) may occur on the stationary phase. This side effect is rather small and exists mainly due to the van der Waals forces or nonpolar interactions with the resin, but these secondary effects usually are negligible.

This technique is applied with equal success to protein purification and in proteomics/peptidomics for peptide separation. In 2D LC-MS/MS approaches, IEC is used as a fractionation step preceding the RP-HPLC step with direct infusion into the mass spectrometer for peptide sequencing. Strong cation exchangers (SCX) are typically used in those cases where there is a possibility to desalt the sample prior to its analysis in a mass spectrometer. Theoretically this approach fits into in-line 2D LC systems where fractions eluted with NaCl from the SCX column are directed to the trapping column for concentration and desalting. Otherwise, volatile buffers would need to be used as a mobile phase. Alternatively, fractions are eluted offline, desalted and then loaded onto a nano-LC C_{18} column interfaced with the mass spectrometer. Although

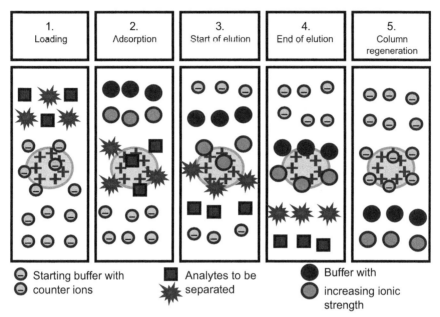

Figure 5.2.1 The principle of IEC with increasing ionic strength elution steps. 1. Starting conditions (column equilibration): Preparation of column for analysis. 2. Adsorption of sample substances on stationary phase in the column. 3. Start of desorption: Usually by changing pH or ionic strength of the mobile phase. 4. End of desorption: To remove from the column all substances not eluted under earlier applied conditions. 5. Regeneration: Column reequilibration with starting conditions and preparation of the column for next experiments.

peptide elution with a pH gradient is possible, it would require a composition of several volatile buffers to cover such a broad pH range on IEC, making this approach impractical. As indicated in multiple articles, offline SCX for fractionation of peptides is now commonly used.

5.2.3 Common Types of Ion Exchange Chromatography Stationary Phases

Each ion exchanger contains charged groups covalently bound to a resin, and this is fundamental for the separation process and determines the type of ion exchange mechanism. These positively or negatively charged groups facilitate electrostatic interactions with the oppositely charged ions, called counter-ions. The counter-ions are reversibly

Table 5.2.1 Common types of ion exchanger functional groups

Functional groups	Cation exchangers		Anion exchangers	
	Strong	**Weak**	**Strong**	**Weak**
	$-SO_3^-$ H^+	$-CO_2^-$ H^+	$\begin{array}{c} CH_3 \\ \| \\ -N^+-CH_3 \quad Cl^- \\ \| \\ CH_3 \end{array}$	$\begin{array}{c} R \\ \| \\ -N^+-H \quad Cl^- \\ \| \\ R \end{array}$

bound to the resin, making it possible for proteins and peptides to bind and elute. Based on this principle of IEC, two systems are available (Table 5.2.1): negatively charged exchangers facilitating exchange cations, called cation exchangers; or positively charged exchangers associated with anions, called anion exchangers. The type of ion exchanger is determined by the type of chemical structure of the functional groups and the capacity. Sulphonic groups and quaternary amino groups are used in strong ion exchangers; other functional groups are used in the weak ion exchangers. For example, ion exchangers containing sulphonic groups are referred to as strong cation exchangers (SCX) and quaternary amino groups are termed strong anion exchangers, respectively. The terms "strong" and "weak" refer to the acid or base strength of the ionic groups attached to the resin. As a result, strong ion exchangers are ionized over a wide range of pH, while for weak ion exchangers the ionization level strongly depends on the degree of dissociation.

The resin in the stationary phase can be based on inorganic compounds (eg, silica); synthetic resins like polystyrenes (eg, styrene-divinylbenzene copolymer, see Fig. 5.2.2); or polysaccharides (eg, cellulose and derivatives). The matrix of a stationary phase can determine separation parameters, such as capacity, efficiency, recovery and selectivity, as well as mechanical properties, chemical stability, and mobile phase flow.

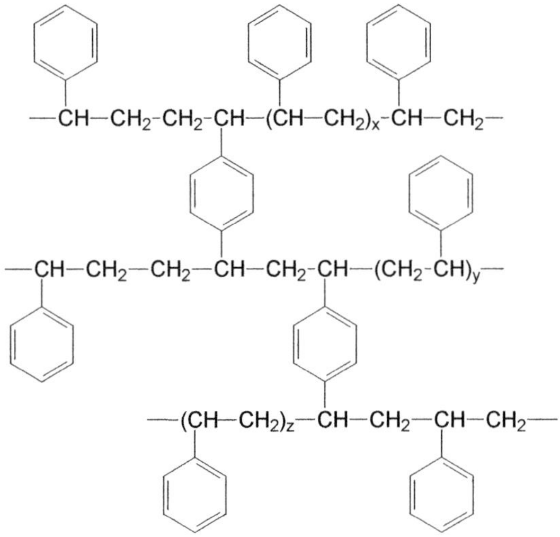

Figure 5.2.2 Structure of cross-linked polystyrene-divinylbenzene polymer.

Early on, cellulose based stationary phases were used for separations of proteins and peptides [2]. This resin was advantageous due to its high hydrophilicity, which limited the tendency to denature proteins. Currently other resins with superior properties of separation are preferred over cellulose because of its low capacity and irregular shape of beads. Initially, the first non-cellulose-based stationary phases for IEC introduced on the market with spherical beads and high rigidity were based on dextran (Sephadex™ [3]) and agarose (Sepharose™). Cross-linked cellulose (DEAE) also became very popular. The disadvantage of these resins is that they can only be used under low pressure, thus low flow. Other stationary phases like highly cross-linked agarose (eg, Sepharose High Performance™, Superose™, and Superdex™) or synthetic polymer matrices (eg, SOURCE™) can operate at higher pressures with both analytical and preparative HPLC systems.

Ion exchange chromatography can also be performed with monolithic columns instead of packed columns. Stationary phases in monolithic columns can be selected as resins in packed columns; however, there are no commercial monolithic IEC columns available on the market. For the separation of proteins monolithic SCX columns have been used with monolithic poly(glycidyl methacrylate-*co*-ethylene dimethacrylate) grafted with poly (2-acrylamido-2-methyl-1-propanesulfonic acid) chains and others [4,5].

5.2.4 Choice of Ion Exchanger (Cation or Anion?)

Any charged molecule can be bound to the stationary phase with the opposite charge. This interaction is facilitated by electrostatic forces and is reversible. For ions with only one type of group (cationic or anionic), the choice of an ion exchanger is clear. Because of the presence of N- and C- terminal ends in polypeptide chains (proteins and peptides) and side chains also containing carboxyl or amine groups, these molecules are amphoteric in nature. As such, proteins and peptides will have an overall positive, neutral or negative charge depending on the pH of the solvent or mobile phase during LC separation. The isoelectric point (pI) is defined as the pH at which the protein/peptide has a net of charge zero. At such a pH, molecules cannot bind to the ion exchange resin.

At pH values lower than its pI, the protein/peptide acquires a net positive charge and thus it will interact with the cation exchanger, while at a pH above the pI, the same protein/peptide will be bound to the anion exchanger due to the net negative charge. The ion exchanger and optimal pH of the buffer are determined by two factors: best separation and protein stability. The primary structures of proteins and peptides are considerably less susceptible to changes of pH compared to their biological activity. For example, most proteins retain their enzymatic activity within a specific pH range. Outside of this range, the enzyme will lose activity, which can be irreversible. An example of protein net charge as a function of pH is shown in Fig. 5.2.3. At a pH lower than 5.0 (below pI), the majority of proteins have a positive net charge and can be adsorbed on the cation exchanger. On the other hand, at a pH higher

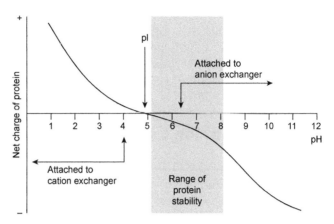

Figure 5.2.3 The net charge of a protein as a function of pH. The range of protein stability is only an example.

than 5.0 (above pI), molecules or proteins have a negative net charge and can be adsorbed on an anion exchanger. Despite the fact that there are two options to separate this one particular protein, only anion exchange chromatography can be used to purify native protein with its retained activity (in this example, it is stable only in the pH range 5.0–8.0).

The majority of proteomic studies performed to date have focused on the quantitative changes of the protein itself and biological activity was of secondary concern. Proteins are denatured, reduced and alkylated for effective enzymatic digestion, which destroys their biological activity. Therefore, considerations described here are important in the case of single protein purification and identification, as well as in proteomic profiling based on tryptic digests of entire samples. As we accumulate global expertise in data interpretation of proteomic profiling, biological activity of proteins begins to play a more important role and it is a huge analytical challenge because of the difficulty of correlation of these two properties of proteins. In the case of enzymes, a sample can be split into two and then one used for activity measurements and the other used for quantitative proteomics. It is more difficult to measure biological function that is associated with structural features of a protein, eg, a functionality of receptors. Several novel approaches were recently developed to meet the demand of measurement of protein expression level and then to match this measurement with protein activity by using interactions with specific substrates or ligands [6].

5.2.5 Choice of Strong or Weak Ion Exchanger

To obtain the best resolution for separation of proteins, the pH of the buffer should be adjusted to the highest or the lowest value at which the protein is still stable and its net charge reaches maximal value (for example, see Fig. 5.2.3). For these conditions, a strong ion exchanger is the best choice. However, for proteins which the pI is between 5.5 and 7.5, weak ion exchangers can be used and can also be applied in those cases where proteins are irreversibly stuck to the column.

Strong ion exchangers are ionized in a wide range of pH, while a number of ionized groups on the weak ion exchangers strongly depend on the pH of the mobile phase. As a result, the capacity of weak ion exchangers is different for various pH values, making it difficult to estimate the maximum amount of the analyte that can be loaded on the column. Moreover, during pH gradient elution, the properties of weak ion exchangers may vary and the entire experiment is uncontrolled. All these features make SCX the best choice for peptide separation. Nevertheless, weak ion exchangers can still be used for other applications, such as separation of native proteins.

5.2.6 Buffers in Ion Exchange Chromatography

As mentioned earlier, the choice of ionic strength and pH of the mobile phase have to be considered. Concentration of the buffer is also important, especially for separation of proteins. Protein structure is stabilized by salts; however, the concentration should be taken into consideration as the protein may precipitate based on the "salt-out" mechanism. There is no one universal concentration of buffer, but

Table 5.2.2 Volatile buffers used in IEC

Buffer substance	pH interval	Counter-ion
Pyridine/formic acid	2.3–3.5	$HCOO^-$
Ammonia/formic acid	7.0–8.5	$HCOO^-$
Ammonia/acetic acid	8.5–10.0	CH_3COO^-
Ammonium bicarbonate	7.9	HCO_3^-
Ammonium carbonate/ammonia	8.0–9.5	CO_3^-

for the majority of separations it should not be lower than 10 mM to have a sufficient buffer capacity. On the other hand, when a sample containing high salt concentration is applied on a column, there is always a possibility that it will not adsorb to the column bead. In that case, it is advisable to dilute the sample or desalt on, for example, a PD-10 column [7] (for intact proteins) or using a Zip-Tip procedure (for peptides), before it is injected into the system.

The most common buffers for ion exchangers are weak buffers (eg, Tris–HCl) together with NaCl gradient or volatile solutions suitable for direct connection to the MS, such as ammonium acetate, ammonium formate, and ammonium bicarbonate. The most common volatile buffers are listed in Table 5.2.2. Phosphates should be avoided due to their conversion to polyphosphates upon high temperature of the heated capillary. This may lead to irreversible clogging of expensive parts of the instrument.

5.2.7 Ion Exchange Chromatography in Proteomic Studies

Rapid development of chromatographic/electrophoretic techniques used in proteomics studies has improved separation of complex mixtures of biomolecules. Multidimensional liquid chromatography (ie, 2D-LC) was developed to increase resolving power and peak capacity for high isoelectric focusing throughput peptide separation. Commonly, IEC is used as a first or middle step of multidimensional separation because of the presence of salts in the mobile phase. The last dimension of the setup usually comprises a reverse phase (RP) chromatography, due to its higher efficiency and online elution using volatile solvents and desalting power prior to mass spectrometry detection.

As discussed earlier, SCX is the most common type of IEC used in high-throughput peptide separation in proteomics. However, this step always requires a desalting step before RP-LC-MS/MS analysis. Here, there are two options. The first involves an *offline approach*, where fractions (usually around 10) collected from the SCX are desalted manually and analyzed in the second dimension [8]. The second approach involves an *online setup*, where fractions from the first column are directly introduced to the second dimension via a trap column (C8, C18) to desalt fractions. Automation of this system is much more complex, including switching valves, and demands some technical skills. The possibility of introducing larger samples during the first dimension is the biggest advantage of offline analysis with a larger column diameter and thus higher capacity. Higher throughput and automation of the entire procedure can easily be achieved by an online analysis. In offline separations, after the SCX stage, each fraction must be manually desalted.

It was shown that SCX separation can be used in analysis targeting peptides with specific functional groups. N-terminal peptides, phosphopeptides, peptides with single and multiple basic residues can be separated by SCX chromatography, and this has become a powerful methodology in peptides analysis. Also 2D-(SCX-RP)-nano-LC-MS/MS with isotope-coded protein label (ICPL™) can be easily applied for quantitative analysis of phosphorylated peptides [9].

New developments of multidimensional chromatography by incorporation of an SCX trap column between the two dimensions of a high/low-pH RP–RP system shows that the new platform offers enhancement in hydrophilic peptide identification [10]. It needs to be acknowledged that increasing the number of dimensions during the separation process is always a cause of errors in quantitative analysis. These arrangements, including eg, several SEC columns linked together, are often efficient but rather complex to reproduce by other laboratories. Although 2D-SCX-RP combination is still the most popular analytical technique in proteome shotgun analysis, there is still an urgent need for a robust, fast, and quantitative combination of orthogonal separations to fulfill present demands.

References

[1] Moore S, Stein WH. Chromatography of amino acids on sulfonated polystyrene resins. J Biol Chem 1951;192(2):663–81.

[2] Sober HA, Peterson EA. Protein chromatography on ion exchange cellulose. Fed Proc 1958;17(4):1116–26.

[3] Novotny J. Chromatography of proteins and peptides on sephadex ion-exchangers: dependence of the resolution on the elution schedule. FEBS Lett 1971;14(1):7–10.

[4] Viklund C, Svec F, Frechet JM, Irgum K. Fast ion-exchange HPLC of proteins using porous poly(glycidyl methacrylate-co-ethylene dimethacrylate) monoliths grafted with poly(2-acrylamido-2-methyl-1-propanesulfonic acid). Biotechnol Prog 1997;13(5):597–600.

[5] Rezeli M, Kilar F, Hjerten S. Monolithic beds of artificial gel antibodies. J Chromatogr A 2006;1109(1):100–2.

[6] Bodzon-Kulakowska A, Suder P, Drabik A, Kotlinska JH, Silberring J. Constant activity of glutamine synthetase after morphine administration versus proteomic results. Anal Bioanal Chem 2010;398(7–8):2939–42.

[7] Tantipaiboonwong P, Sinchaikul S, Sriyam S, Phutrakul S, Chen ST. Different techniques for urinary protein analysis of normal and lung cancer patients. Proteomics 2005;5(4):1140–9.

[8] Barbhuiya MA, Sahasrabuddhe NA, Pinto SM, Muthusamy B, Singh TD, Nanjappa V, et al. Comprehensive proteomic analysis of human bile. Proteomics 2011;11(23):4443–53.

[9] Fleron M, Greffe Y, Musmeci D, Massart AC, Hennequiere V, Mazzucchelli G, et al. Novel post-digest isotope coded protein labeling method for phospho- and glycoproteome analysis. J Proteomics 2010;73(10):1986–2005.

[10] Kong RP, Siu SO, Lee SS, Lo C, Chu IK. Development of online high-/low-pH reversed-phase-reversed-phase two-dimensional liquid chromatography for shotgun proteomics: a reversed-phase-strong cation exchange-reversed-phase approach. J Chromatogr A 2011;1218(23):3681–8.

5.3

PROTEIN AND PEPTIDE SEPARATION BASED ON ISOELECTRIC POINT

A. Drabik
AGH University of Science and Technology, Krakow, Poland

J. Silberring
AGH University of Science and Technology, Krakow, Poland;
Polish Academy of Sciences, Zabrze, Poland

5.3.1 Principles of Isoelectric Focusing

Isoelectric focusing (IEF), introduced in 1967 [1], is based on enrichment of proteins and peptides due to their ability to act like acids or bases, depending on the pH environment. Non-amphoteric species, such as nucleic acids, cannot be resolved by focusing. IEF concentrates amphoteric molecules in segments that are created by specific pH ranges, which correspond to the isoelectric points (pI) of proteins and peptides (Fig. 5.3.1). Advantages of IEF include high resolving power during fractionation while simultaneously being able to concentrate the sample.

Proteins and peptides are concentrated by utilizing the charge located on proteins due to post-translational modifications, prosthetic groups, and side chains of amino acids. Peptides and proteins gain positive charge in low pH environments and a negative charge in basic solutions. The total charge of the molecule is the algebraic sum of all negative and positive charges, which influences the direction of migration. Proteins and peptides localized at a pH below the pI move toward the cathode (negative), and when localized in a pH above the pI, the molecule

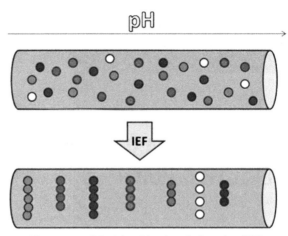

Figure 5.3.1 Principle of isoelectric focusing (IEF).

migrates toward the anode. Proteins and peptides gain or lose protons during the migration. When a molecule reaches the pI value, migration ceases and the molecule achieves equilibrium, thereby starting to concentrate at this point. If the particle diffuses to a region of higher pH, protonation occurs and the molecule is forced by the electric current to move to the cathode. It is also worth noting that some charges might be "hidden" inside the protein structure and do not contribute to the net charge of the molecule [2].

Amphoteric compounds can be focused within a very narrow pH range, while low molecular weight species cannot be resolved using the IEF technique because of a massive diffusion effect, which causes blurring of dissolved fractions. It is preferable to focus particles with molecular weight larger than 2 kDa, when diffusion is less pronounced. This diffusion could cause spreading of the analytes to the neighboring compartments, decreasing sensitivity and resolution.

IEF resolution is dependent on the pH range used during the focusing process, type of buffering agent, time and applied electric current. Usually the maximal resolution of this technique reaches 0.02–0.001 pH unit. Application of zoomed gradients, in a very narrow range (for example 3.5–4.5 pH) can also cause reduction of resolving power as

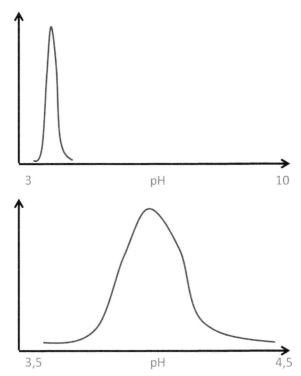

Figure 5.3.2 Peak broadening associated with narrowing pH gradient in IEF.

illustrated by peak broadening in Fig. 5.3.2. This effect is typically observed when proteins or peptides are differentially charged as a result of interactions with sample contaminants. Buffers with very low conductivity allow for the highest possible voltage gradients, which leads to a much improved resolution during separation of peptides and proteins due to reduced analysis times.

The IEF process should be performed with the highest possible electric current. The initial voltage values will be high until the ampholytes create a uniformly increasing gradient; then during the focusing process the voltage values will decrease. IEF is performed with constant power to prevent the system from overheating. Once the power supply is off, the gradient starts to blur and the focused molecules begin to diffuse. To minimize such a situation, focusing is often processed in sucrose or glycerol solutions. Application of different pH gradient carriers is frequently recommended to increase the reproducibility, resolution, and handling of the focusing procedure. For proteomic purposes, the most common media are polyacrylamide and agarose gels (immobilized gradient), as well as chambers separated with semipermeable membranes (in solution) [3].

5.3.2 Sample Preparation Prior to Isoelectric Focusing

Materials suitable for IEF include whole-cell lysates and semipurified or immunoprecipitated proteins. It is important to avoid conditions that can cause chemical modifications to the proteins, such as high urea concentrations together with high temperatures (around 60°C), which may cause modifications at N-termini. Charge changes and modifications can cause a protein to migrate as two or more bands, resulting in a sensitivity decrease. Complete reduction and alkylation of disulfide bonds is necessary to limit the number of modified species.

One of the most severe drawbacks of all IEF techniques is protein precipitation at their pI value. This problem can be solved by increasing the sample concentration or by using 8 M urea. Overloading is often necessary in order to reveal minor components by decreasing the ionic strength. Under unfavorable conditions, like high temperature or high salt concentration, protein precipitation will not occur at a precise pH point but will be detected as smears covering as much as a 0.5 pH unit. Glycerol, ethylene and propylene glycols are used with success for protein dissolution. For example, addition of 30% glycerol stabilizes and solubilizes proteins and prevents protein precipitation as a result of Joule's heating, a warming effect due to the electrical current. There are still a number of proteins completely insensitive to these solubilizers. Non-denaturing detergents, zwitterions or taurine, 20% sucrose, or sorbitol can be also adopted for this purpose [4].

Addition of CHAPS, CHAPSO, NP-40 or Triton X-100 (nonionic surfactants) allows for highly hydrophobic (cytosolic and membrane) protein solubilization. Furthermore, hydrophobic proteins require the presence of 8 M urea to stay in solution. High urea content leads to conformational changes in many proteins, and disruption of their quaternary structure (caused by hydrogen bond disruption), thereby avoiding unwanted aggregation. Moreover, supplementation with EDTA may prevent formation of water-soluble conglomerates through chelation.

Desalting prior to IEF is a necessity as protein precipitation, localized concentration, and denaturation can also be induced by the presence of high salt levels in the sample. Proper gradient separation will not occur in the presence of high salt. Usually, uneven conductivity profile characteristic for the extreme values of pH occurs around neutral pH regions. This is why the proteins with the most common pI will not separate properly with high electric current applied in the presence of salt.

Biological contaminants such as nucleic acids in the sample can also interfere with IEF, creating streaks. To avoid this, samples should be treated with RNAse and DNAse. Alternatively, an isopropanol/isobutanol mixture can precipitate nucleic acids.

Because it is possible to subject both proteins and peptides to IEF, it has to be taken into consideration what direction is the better solution for achieving the experimental goals. IEF of proteins may be preferred when there is concern that a possible diffusion effect of smaller molecules like peptides might alter experimental results. However, in some cases the use of peptides is more beneficial, especially when analyzing very alkaline or acidic proteins. Proteins that are localized in very extreme fractions are not focused favorably due to Joule's heating effect and thereby causing proteins to precipitate. Finally, using a protein digest is beneficial, because after enzymatic fragmentation it is more likely that the resulting peptides will possess a more evenly distributed pI, not just the marginal pI of their parent protein. This may lead to enhanced protein identification by MS [5].

5.3.3 Isoelectric Focusing in Liquid State

IEF in solution is performed in specially designed chambers consisting of compartments separated by semipermeable membranes to minimize diffusion of the molecules between cells. Such partition offers resistance to fluid convection, but does not disturb the flow of proteins between cells. Chambers are localized horizontally with rotation, to minimize the effect of gravitation. Focusing is achieved in the presence of chaotropic substances in high concentration, such as urea, thiourea, and nonionic detergents to disrupt the hydrophobic interactions between proteins, and to counteract their precipitation. Laboratory-scale focusing is accomplished in Rotofor Cell devices (Bio-Rad, USA) or ZOOM IEF (Invitrogen, USA), where the primary difficulty is sample loading and care must be taken to load the sample without any air bubbles in the focusing chamber. Typically, IEF is carried out in solution for preparative purposes like fractionation of proteins in complex mixtures.

The most popular approach to create pH segments is to use carrier ampholytes. These are compounds that contain both acidic and basic groups and are capable of generating a regularly increasing pH gradient due to the applied electric current. Usually, the mixture of ampholytes consists of hundreds or thousands of oligomers.

Another technique of generating the pH gradient is by utilizing ion pairs of specifically selected buffers. For example MPOS (3-(N-morpholino)propanesulfonic acid) is the buffering agent that titrates N-aminobutyric acid in a pH range between 4.9 and 6.2. Similarly, MES (2-(N-morpholino)ethanesulfonic acid) is an ion pair of Gly–Gly peptide that works in the pH range 4.9–5.6. Ion pairs must be comprised of compounds that are characterized by insignificant electrophoretic mobility and low affinity interactions with the separated proteins and peptides. This application of IEF is, however, reserved for large samples. Usually, focusing chambers have dimensions of a couple of milliliters.

5.3.4 Immobilized pH Gradient Isoelectric Focusing

Focusing using immobilized polyacrylamide gel (IPG) strips is typically applied as a first step of two-dimensional gel electrophoresis. Free radicals are formed by the use of a combination of riboflavin, ammonium persulfate (APS) and tetramethylenediamine (TEMED). Riboflavin is a photoinitiator and light generates free radicals, whereas APS chemically decomposes to its sulfate radicals. TEMED serves as the free base, acting as an accelerator for ammonium persulfate and riboflavin decomposition. The combination of the two reaction initiators results in more complete polymerization in gels containing low pH-ampholytes than does chemical polymerization alone. Combination of IEF and sodium dodecyl sulfate polyacrylamide gel electrophoresis (SDS-PAGE) allows for imaging the pI and molecular size in a single experiment, making visualization of posttranslational modifications possible. Usually, precast IEF gels are used and have become more popular because of their reproducibility and high resolution.

Agarose is a natural polysaccharide, which consists of long-chain, complex-sugar molecules cross-linked by hydrogen bonds. Large-molecular-weight proteins have a limited migration in polyacrylamide gels and will move through agarose due to the greater pore size. Therefore, this type of gel is applied for proteins greater than 200 kDa, because electro-osmosis disturbs the focusing process of smaller species [6].

5.3.5 Capillary Isoelectric Focusing

Capillary IEF (CIEF) is analogous to conventional IEF; however, the separation is performed in fused silica capillaries with an internal diameter of 25–100 μm. The principle of CIEF is similar to that of a gel, where proteins migrate within a stable pH gradient formed by carrier ampholytes under the control of an electric field. At equilibrium, proteins become focused within the pH gradient where they have a balanced net charge. Any diffusion of the focused protein away from its isoelectric zone will result in gaining of a charge, resulting in back migration to the sector. This approach differs from the previously described techniques in that focused zones must be transported past the monitoring point to detect the separated proteins. The use of very narrow fused silica capillary as the separation chamber provides efficient dissipation of heat, allowing for the use of high voltage. The major advantage of using a short capillary tube or a microchip channel is the shorter separation time. The focusing time is expected to be proportional to the length of a capillary and to the field strength. CIEF has been used successfully for the characterization of proteins with very delicate differences in structure, especially in the biopharmaceutical industry.

5.3.6 Isoelectric Focusing in Living Organisms

The isoelectric focusing mechanism is also adopted by living cells for the purpose of providing the proper environment for biochemical reactions. This purely biological system can cause a few orders of magnitude difference in enzyme concentrations between cell compartments. In 1984, Slavik and Kotyk demonstrated the presence of a continuous pH gradient ranging from pH 7.2 in the center of the cell to 6.4 in the cell periphery. Because the efficiency of IEF is directly proportional to electric field strength and pH gradient, it can be calculated that the resolution in living cells is more than four orders of magnitude higher than in any manmade apparatus [7].

The large diversity in pI values among natural proteins suggest that this parameter has biological meaning and that it is controlled by natural selection. Studies show that this parameter is controlling the entry of a protein into the nucleus. Posttranslational modifications like phosphorylation or dephosphorylation regulate a molecule's location within the cell, where they can act as regulators of diverse biochemical functions.

5.3.7 Summary

IEF is a high-resolution technique that can resolve proteins differing in pI by less than 0.05 pH unit and is generally carried out under nondenaturing conditions, in which antibodies, antigens, and enzymes maintain most of their biochemical properties. Because of the complexity of the manufacturing process, including carrier ampholytes, IEF exhibits some batch-to-batch variability and high reproducibility in peak position can be expected. When precise pI positioning is necessary and reproducible patterns are required, focusing should be performed in an immobilized gradient, while for fractionation purposes the IEF is carried out in solution. IEF allows for efficient protein and peptide fractionation that is necessary for further proteomic investigation in complex biological samples.

References

[1] Haglund H. Isoelectric focusing in pH gradients—a technique for fractionation and characterization of ampholytes. Methods Biochem Anal 1971;19:1—104.
[2] Bierczyńska-Krzysik A, Lubec G. Two dimensional gel electrophoresis. Hoboken (NJ, USA): John Wiley & Sons, Inc.; 2008.
[3] Sounart TL, Safier PA, Baygents JC. Theory and simulation of isoelectric focusing. Amsterdam (The Netherlands): Elsevier Inc.; 2005.
[4] Berkelman T. Generation of pH gradients. Amsterdam (The Netherlands): Elsevier Inc.; 2005.
[5] Westermeier R. Slab gel IEF. Amsterdam (The Netherlands): Elsevier Inc.; 2005.
[6] Herbert B. Some practices and pitfalls of sample preparation for isoelectric focusing in proteomics. Amsterdam (The Netherlands): Elsevier Inc.; 2005.
[7] Righetti PG. Theory and fundamental aspects of isoelectric focusing. (USA): Elsevier Biomedical; 1983.

5.4

CAPILLARY COLUMNS FOR PROTEOMIC ANALYSES

M. Smoluch
AGH University of Science and Technology, Krakow, Poland

J. Silberring
*AGH University of Science and Technology, Krakow, Poland;
Polish Academy of Sciences, Zabrze, Poland*

5.4.1 Introduction

Liquid chromatography has become the most powerful technique in proteomic separation science due to continuous technological improvement of capillary column production, implementation of modern stationary phases, and development of LC methods in conjunction with mass spectrometry. However, it is important to note that most of the progress made has been in the area of peptide rather than protein separation. Gel electrophoresis remains a powerful technique for intact protein separation and is quite often used. Otherwise, proteins are fragmented by enzymatic digestion and resulting peptides provide the basis for protein identification and quantitation. Depending on sample complexity, one- or multidimensional LC is applied. Almost exclusively, capillary columns (ID < 0.3 mm) are interfaced within a nanospray source because they require much less sample for successful analysis with substantially increased sensitivity. Recently, LC proteomics has been directing its analytical efforts toward using monolithic columns, which are becoming competitive with the columns containing standard stationary phase packings. In this section, capillary columns for proteomic analyses are described including conventional columns and monoliths.

5.4.2 Conventional Capillary Columns

During the last few years, monolithic columns have gained popularity in proteomics, but this does not make conventional columns packed with porous materials obsolete. Development of the latter types of columns is progressing by production of novel, modified stationary phases with very small particle sizes (<3 μm). Routinely applied flow rates used in conventional columns are shown in Table 5.4.1 and a typical protein tryptic digest separation is presented in Fig. 5.4.1. The appropriate selection of the flow rate is important for efficient separation and is dependent on column dimensions, stationary phase particle size, system temperature, and solvents used.

Shorter columns with larger internal diameter (I.D.) produce lower back pressure, allowing peptides to be eluted at higher flow rates. In general, the larger the particle size, the lower the system back pressure for a given flow rate. Smaller particles generally provide greater surface area for the same volume and better separation but also create higher system back pressure. Columns packed with very small particle sizes usually require very high back

Table 5.4.1 Column diameter versus flow rate used

Column internal diameter (mm)	Typical flow rate (μl/min)
4.6	1000
2.1	200
1.0	50
0.3	4
0.075	0.3
0.05	0.1

Values are given for columns with particle size 5 μm, length of 25 cm and separation performed at room temperature. Columns with diameter smaller than 0.3 mm considered as capillary columns.

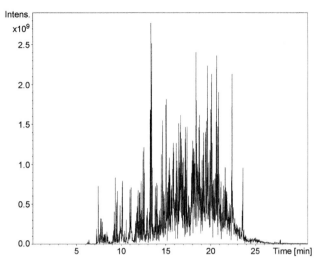

Figure 5.4.1 Total ion chromatogram of human serum albumin tryptic digest performed on a C18 RP column (10 cm × 75 μm, 3-μm particle size) at a flow rate of 0.3 μl/min and gradient elution from 0% to 80% acetonitrile with 0.1% formic acid in 25 min.

pressures and as a consequence, a dedicated LC system is required. Such systems are referred to as UPLC (ultra-performance liquid chromatography), although this term is a trademark of Waters Corp. In HPLC the maximal system back pressure does not exceed 40 Mpa (5800 psi), while in UPLC can reach 100 MPa (14,500 psi). This increase in tolerated pressure permits use of smaller particle sizes (>2 μm) and offers better resolution of peptide peaks and much shorter analysis times. To fully utilize the advantage of UPLC, a very fast detector (mass spectrometer) is required, as the typical peak width is only 3–5 s. This has to be sufficient to collect full scan and data dependent MS/MS spectra to identify compounds in the mixture (usually protein tryptic digest). UPLC in small capillaries (nano UPLC) remains challenging due to difficulties in packing long capillaries with small particles, although it is only a matter of time before this will become the standard approach. Table 5.4.2 compares capillary HPLC and UPLC when both are working close to system upper limits [1].

Table 5.4.2 HPLC versus UPLC

75 μm ID C18 column	
HPLC (35 MPa)	UPLC (80 MPa)
50 cm; 3 μm	50 cm; 2 μm
	100 cm; 3 μm
	250 cm; 5 μm

Longer column with the same particle size, or column with the same length, but with smaller particles provides better resolution.

5.4.3 Monoliths

A monolith can be defined as a continuous porous object whose morphology and pore structure can be modified in a wide range [2]. The structure of a monolith contains channels instead of beads. In general they are prepared by polymerization of monomers with inorganic- or organic-based skeletons. The reaction is performed directly in the fused-silica capillary, with the inner surface functionalized with vinyl groups in porogenic solvent, in which the monomers but not the polymers are soluble. The resulting monolithic polymer bead is a uniformly porous core integrated with the capillary wall. The strength of this structure is very high, with no need for frits or encapsulation. The latter is an advantage because even a broken column retains the stationary phase and it can still be used. A typical monolith is depicted in Fig. 5.4.2. Depending on the type of monomer used, chemistry of synthesis, and surface reactions, the monomers can gain different functionality.

5.4.3.1 Silica-Based Monoliths

Silica-based monolithic columns were the first monolithic columns introduced to the market. They were prepared using sol–gel technology, which enables production of a continuous sol–gel network

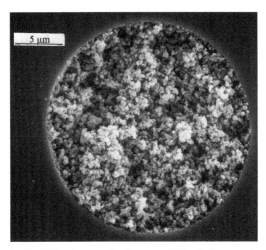

Figure 5.4.2 SEM micrograph of a monolithic stationary phase packed in 25 μm ID fused silica. http://www.dac.neu.edu/barnett/ KargerRG/Jian_monolith.htm, courtesy of Barry L. Karger, Northeastern University, Boston.

in the capillary [2]. Another method uses this technology to glue silica-based, conventionally packed particles in order to produce a continuously bonded bead. For this type of stationary phase, the back pressure was low due to the high porosity of monolith. As a consequence, protein digests could be analyzed using high flow rates and long columns, thus improving resolution. It is worth noting that for all silica-based columns, the pH range of mobile phase is restricted. Such columns cannot be used at high pH values but fortunately in proteomics usually acidic conditions in a gradient mode are required for a typical LC MS/MS run. Silica-based monoliths have several strong advantages as they are mechanically stable, possess higher loading capacity, and are resistant to swelling or shrinking caused by rapid changes of the eluent. They are characterized by very good resolution and reproducibility in addition to the known advantages of excellent column pressure stability over a time period of 6 months.

Depending on their application, columns may differ in dimensions, flow rate, and total run times applied for separation. In many cases a precolumn is also used prior to the analytical column. This is

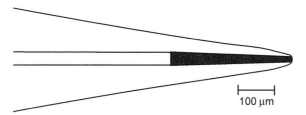

Figure 5.4.3 Schematic diagram of a nanoelectrospray emitter packed with stationary phase. Typical column ID is 50 μm (approx. 10−15 μm on the tip). Bed length varies from 10 to 250 μm.

particularly important in the case of capillary columns with an I.D. below 75 mm, as even the smallest particles from solvents or samples can irreversibly block the column. In case of monolithic packings, this clogged piece can be cut away and as no frit is present, the remaining column length can still be used. Moreover, the precolumn (or trap column) serves to desalt fractions during 2D runs. To improve total performance of LC-MS/MS systems, integrated nanoelectrospray emitters are developed (see Fig. 5.4.3.).

5.4.3.2 Organic-Based Monoliths

The second type of monolith is organic polymer-based, which is formed inside the capillary by a polymerization chain reaction. Synthesis can be acrylamide-based, methacrylate-based or styrene-based, and the reaction mixtures usually consist of a combination of monomers and a cross-linker, initiator, and a porogenic mixture of solvents (see Fig. 5.4.4.). Depending on the monomer used for production, the monolith will gain specific functionality, such as hydrophobicity or hydrophilicity. The overall porosity of the monolith can be controlled by adjusting the concentration of a cross-linker applied in the reaction. An initiator is necessary to start the step-wise chain reaction with the use of heat or UV

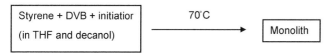

Figure 5.4.4 Schematic diagram of monolith preparation.

light, and column functionalities can be achieved by using specific chemistries and precursors. Detailed description of monolithic polymer column preparation can be found elsewhere [3,4].

5.4.3.3 Methacrylate-Based Monoliths

By modifying the reactive epoxide group, functionalities can be produced to create columns such as strong cation exchange (SCX), anion exchange (AX) or reverse phase (RP) columns. The most frequently used monomer in methacrylate-based monoliths is glycidyl methacrylate (GMA); however, methacrylate capillary monoliths are not produced commercially, therefore only homemade columns are available. Similarly to the silica-based monoliths, methacrylates also generate low back pressure, and as a consequence, high flow rates or longer columns are used. Methacrylate materials were reported to play a role in online enzyme reactors [5] used for affinity chromatography, where intact proteins were attached to the column material [6]; or for typical separation of protein digest. In general, methacrylate columns may be produced for many purposes as diverse groups can be attached to the monolith material during the production process. Comparison of the silica-based and organic-based monoliths is shown in Table 5.4.3.

5.4.3.4 Styrene-Based Monoliths

The PS-DVB (polystyrene-divinylbenzene) stationary phase has hydrophobic properties and can be directly used for RP separations without any modifications, and is comparable with C4 and C8 reverse-phase columns. This type of column is commercially available with lengths of 5 and 25 cm, and IDs of 100 and 200 μm. Typical flow rates for such columns are in the range of 1−3 μl/min [8], with total analysis times of approximately 20 min. If necessary, a styrene-based precolumn (offline or online) can be mounted prior to the analytical column. This type of monolithic column can also be used for direct analysis of intact proteins (top-down approach). Styrene-based monolith columns

Table 5.4.3 Comparison of pore structural and operational parameters of monolithic polymeric columns and research monolithic silica columns

Property	Silica monoliths	Organic-based monoliths
Pore modality	Bimodal (macro + meso)	Unimodal (broad)
Macropore diameter	1–10 μm	0.05–10.0 μm
Macroporosity	40–70%	10–95%
Total column porosity	0.8–0.9	Low-density and high density monoliths
Pore morphology	Spongy and worm-like structure	Globular structure
Surface functionality	Introduced by grafting from and grafting on surface modification	Adjusted by functional co-monomers
Column format	Analytical to capillary formats	Preparative to capillary size
Column pressure drop	1/3–1/5 as compared to 5-μm packed columns	Distinct higher column back pressure as compared to monolithic silica columns
Linear velocity range	1–7 mm/s and higher	1–7 mm/s and higher
Plate height	$H = 5$–10 μm at optimum u (*linear velocity of the eluent*)	$H = 5$–10 μm at optimum u
pH range for application	Acidic to pH 8	Acidic to strong alkaline
Typical application areas	Separation of low-molecular-weight compounds and peptides	Separation of peptides and proteins (analytical), isolation and purification of biopolymers

Adapted from Unger KK, Skudas R, Schulte MM. Particle packed columns and monolithic columns in high-performance liquid chromatography-comparison and critical appraisal. J Chromatogr A 2008; 1184(1–2):393–415. Epub 2008/01/08.

generally can be divided into continuous-bead or porous-layer open-tubular columns.

The production of chromatographic columns has been markedly simplified by the concept of continuous-bed chromatography [9]. As stated, the bed is a continuous rod prepared by polymerization of water-soluble monomers directly in a chromatographic tube. This type of column shows enhanced performance at elevated temperatures (60−70°C). Such conditions allow for reduction of back pressure, thereby increasing the resolution of separation and remain safe for the bed of the column. In the majority of proteomics applications, formic acid is used as mobile-phase additive, although there are some studies where trifluoroacetic acid (TFA) is preferred. It has to be noted that selection of the mobile-phase additives is a compromise between chromatographic and mass spectrometry perfor-mance. TFA is known to suppress ion abundance but works efficiently as an ion-pairing agent in the silica-based phases that are not fully end-capped. Porous-layer open-tubular (PLOT) columns comprise of a second type of styrene-based column and are analogous to typical columns used in gas chromatography, where only the wall of the column is coated and not the entire space inside (open tubular columns). Columns are prepared by one-step styrene and divinylbenzene polymerization [10]. As a consequence of the geometry of the open-tubular column, the system works under extremely low back pressure. This permits the use of columns even a few meters long, using low flow rates and extended analysis times to result in a higher number of proteins being identified. In the first application re-ported [10], a 4.2-m long column was used with I.D. of 10 μm and a flow rate of 20 nl/min. Here 1793 peptides (512 proteins) from *Methanosarcina aceti-vorans* were identified in 3.5 h. In another approach, the same research group [11] identified 1209 proteins from the cervical cancer cell line using an additional triphasic trapping column (RP/SCX/SPE) in front of an analytical PLOT column.

5.4.4 Summary and Conclusions

The decision of what type of column should be used in proteomic analysis depends on the type of application (see Table 5.4.4). In principle, for

Table 5.4.4 Separation approach selection dependent on application

Application/ method	Few proteins in a digest	Complex protein digest	Very complex protein digest	Mixture of intact proteins	High- throughput analyses
Conventional 1D LC	+	+?	−	−?	−
Conventional 2D LC	−	−?	+	−	−
Monoliths, silica-based	+	+	+	+?	+
Monoliths, organic-based	+	+	+ (PLOT)	+?	+

Conventional columns include those packed with standard porous materials.

a simple approach, such as separation of a few digested proteins, all of the above described columns can be successfully used. This also applies for conventional columns packed with porous materials. The more complex the sample, the more important it becomes to select the appropriate column. For high throughput and fast analyses the best choice seems to be silica-based monoliths at high flow rates. For extremely complex samples, separation in one dimension may not be sufficient and thus a second dimension must be implemented in the method.

Monolithic columns have a huge potential due to their exceptional chromatographic resolution and the possibility of operating at low nanoliter flow rates, which enhances ionization efficiency. Another advantage of monolithic columns is that they are cost-efficient, because monolithic stationary phases can easily be produced in the laboratory without a need to purchase expensive media and packing devices, and production protocols can be easily found in research literature [4].

5.4.5 Recent Developments

Since the first edition of this book, several devel opments in the field of capillary columns have been achieved.

Fast and efficient HPLC analyses are required in the pharmaceutical industry. This can be achieved by the automation of an entire analytical process, and by minimizing use of reagents. To emphasize this point, a simple example will be given. Standard, analytical reversed-phase columns (4.6 mm × 15 cm) require elution at a flow-rate of 1 ml/min, which utilizes ~30 ml of solvents in a single run. Capillary columns (I.D. 75 µm, 1 cm) are eluted at a flow-rate of 300 nl/min and need ~15 µl for a complete separation. This also has a strong impact on environmental protection and total cost of analysis.

To satisfy even more demanding requirements, micro-total analysis systems (microTAS), also called a lab-on-a-chip, can be used. Such units accomplish sample preparation, separation, and detection, and those processes are integrated in a single and disposable system. Sample preparation steps may include dialysis, preconcentration, extraction, and derivatization of the analyte, followed by their separation in the form of either chromatography or electrophoresis. Laser-induced fluorescence microscopy or mass spectrometry serve as sensitive detectors. The main drawback of such devices is high cost, which in general limits their usage. It has to be noted that clogging of the transfer lines or column on the chip may lead to the replacement of an entire device. The technology of manufacturing microTAS devices is continuously improving, which makes these products less expensive but still beyond the financial recourses of many research laboratories.

Recently, a novel plug-and-play LC-MS source has been developed [12]. This source enables an automated connection between the capillary trap column, separation column, and the electrospray emitter. It can be applied for fast connections in ultra-high-pressure systems, and can be used with systems from any vendor as well as homemade capillary columns. Each component of the source can be easily replaced, if necessary, without interfering with other elements. This system is gaining interest as it can make proteomics analyses much easier. Even for an experienced user, nanoflow separations can cause technical problems, mainly related to incorrectly mounted fittings. This leads to leaks or dead volumes in the system, which usually ruins separations. Plug-and-play systems open a new and attractive field also for less experienced users. This solution does not guarantee highly reproducible separations itself. Good-quality columns are also crucial. For this purpose, plug-and-use fritting technology has been developed [13]. This technology guarantees excellent column performance in terms of retention time, peak width, and peak capacity for separations of complex protein digests. A plug-and-use approach can be especially useful when performing high-throughput separations.

For specific applications, the proper column has to be selected. The criteria for choosing the appropriate column are mainly dependent on the sample

Table 5.4.5 Column selection for low-molecular-weight samples (M_w < 5000 Da)

Sample type		Separation mode
Organic soluble	Hexane soluble	Normal phase adsorption
		Normal phase bonded
	MeOH, MeOH/H$_2$O soluble	Reversed phase bonded
		Chiral
	THF soluble	Gel permeation
Aqueous soluble	Non ionic	Reversed phase
		Chiral
	Ionic	Ion paring/reversed phase
		Ion-exchange
		HILIC
		Chiral
	Peptides	Reversed phase

Table 5.4.6 Column selection for high-molecular-weight samples (M_w > 5000 Da)

Sample type	Separation mode
Organic soluble	Gel permeation
Aqueous soluble	Gel filtration
	Ion-exchange
	Reversed phase
	Hydrophobic interaction
	Affinity/bioaffinity

type, its molecular weight and solubility. The basic guide for LC column selection is shown in Tables 5.4.5 and 5.4.6.

References

[1] Sneekes E-J, Dekker K, de Haan B, Swart R. Optimizing particle size and column length: what is the best way to utilize nano UHPLC in proteomics? [Technical note]. 2011. www.thermoscientific.com/dionex. Available from: http://www.dionex.com/en-us/webdocs/110873-PO-HPLC-OptimizeSizeLength-Proteomics-02Nov2011-LPN2940-01.pdf.

[2] Unger KK, Tanaka N, Machtejevas E. Monolithic silicas in separation science: concepts, syntheses, characterization, modeling and applications. John Wiley & Sons; 2011.

[3] Nischang I, Brueggemann O, Svec F. Advances in the preparation of porous polymer monoliths in capillaries and microfluidic chips with focus on morphological aspects. Anal Bioanal Chem 2010;397(3):953−60.

[4] Rohr T, Hilder E, Donovan J, Svec F, Fréchet J. Photografting and the control of surface chemistry in three-dimensional porous polymer monoliths. Macromolecules 2003;35(5):1677−84.

[5] Lin W, Skinner CD. Design and optimization of porous polymer enzymatic digestors for proteomics. J Sep Sci 2009;32(15−16):2642−52.

[6] Bedair M, El Rassi Z. Affinity chromatography with monolithic capillary columns I. Polymethacrylate monoliths with immobilized mannan for the separation of mannose-binding proteins by capillary electrochromatography and nano-scale liquid chromatography. J Chromatogr A 2004;1044(1−2):177−86.

[7] Unger KK, Skudas R, Schulte MM. Particle packed columns and monolithic columns in high-performance liquid chromatography-comparison and critical appraisal. J Chromatogr A 2008;1184(1−2):393−415. Epub 2008/01/08.

[8] Marcus K, Schafer H, Klaus S, Bunse C, Swart R, Meyer HE. A new fast method for nanoLC-MALDI-TOF/TOF-MS analysis using monolithic columns for peptide preconcentration and separation in proteomic studies. J Proteome Res 2007;6(2):636−43. Epub 2007/02/03.

[9] Hjertén S, Liao J, Zhang R. High-performance liquid chromatography on continuous polymer beds. J Chromatogr A 1989;473:273−5.

[10] Yue G, Luo Q, Zhang J, Wu SL, Karger BL. Ultratrace LC/MS proteomic analysis using 10-microm-i.d. Porous layer open tubular poly(styrene-divinylbenzene) capillary columns. Anal Chem 2007;79(3):938−46. Epub 2007/02/01.

[11] Luo Q, Yue G, Valaskovic GA, Gu Y, Wu SL, Karger BL. On-line 1D and 2D porous layer open tubular/LC-ESI-MS using 10-microm-i.d. poly(styrene-divinylbenzene) columns for ultrasensitive proteomic analysis. Anal Chem 2007;79(16):6174−81. Epub 2007/07/13.

[12] Bereman MS, Hsieh EJ, Corso TN, Van Pelt CK, MacCoss MJ. Development and characterization of a novel plug and play liquid chromatography-mass spectrometry (LC-MS) source that automates connections between the capillary trap, column, and emitter. Mol Cell Proteomics 2013;12:1701−8. Epub 2013/02/19.

[13] Xiao Z, Wang L, Liu Y, Wang Q, Zhang B. A "plug-and-use" approach towards facile fabrication of capillary columns for high performance nanoflow liquid chromatography. J Chromatogr A 2014;1325:109−14. Epub 2013/12/11.

6

IMMUNOAFFINITY DEPLETION OF HIGHLY ABUNDANT PROTEINS FOR PROTEOMIC SAMPLE PREPARATION

J. Wiederin and P. Ciborowski

University of Nebraska Medical Center, Omaha, NE, United States

CHAPTER OUTLINE

6.1 Introduction   101
6.2 Immunodepletion Techniques   102
6.3 Capacity of Immunodepletion Columns
    and Other Devices   104
6.4 Reproducibility   105
6.5 Quality Control of Immunodepletion   106
6.6 Albuminome   107
6.7 Summary   113
References   113

6.1 Introduction

The HUPO Plasma Proteome Project (HPPP) 2005 multicenter study reported that MS–MS datasets from all participating laboratories identified 15,710 proteins based on the International Protein Index (IPI). After applying an integration algorithm to multiple matches of peptide sequences, this dataset yielded 9504 proteins based on IPI and identified with one or more peptides. Of these 9504 proteins, 3020 proteins were identified with two or more

Proteomic Profiling and Analytical Chemistry. http://dx.doi.org/10.1016/B978-0-444-63688-1.00006-9

peptides and constituted the core dataset [1]. In 2003, Anderson and Anderson published a comprehensive overview of the plasma proteome showing a 10^{12} range of protein concentrations in plasma with hemoglobin, albumin and immunoglobulins as the most abundant and interleukins as the least abundant proteins [2]. Considering these two reports and the fact that there is no platform able to analyze proteins (peptides) with such a wide range of concentrations, the need for removal of the most abundant proteins became obvious.

6.2 Immunodepletion Techniques

The necessity to immunodeplete plasma prior to proteomic analysis prompted the question of how many and which proteins should be removed to narrow the concentration range enough to successfully measure and quantitate all remaining proteins. It was obvious that plasma must be free of hemoglobin and should be depleted of albumin and immunoglobulins. This led to the development of liquid chromatography (LC) columns and spin devices based on affinity or immunoaffinity principles to remove the abundant protein(s). These methods, starting with the removal of the two most abundant proteins—albumin and IgG—as well as spin and high-capacity LC columns, evolved over time and now are able to remove many proteins as well as have increasing capacity. In 2005, Bjorhal and coauthors performed systematic and formal comparison of the following devices: Aurum Serum Protein minikit from Bio-Rad, Albumin/IgG Removal Kit from Merck Biosciences, Multiple Affinity Removal System from Agilent Technologies, POROS® Affinity Depletion cartridges (Anti-HSA and Protein G) from Applied Biosystems, and Albumin and IgG Removal Kit from Amersham Biosciences (currently discontinued product) [3]. These five depletion columns each removed a minimum of 94% IgG and 96% albumin from serum. The authors reported that the Multiple Affinity Removal System from Agilent depletion of the six most abundant proteins removed up to ~87% of total protein content in serum, reducing the number of proteins from 3020 to 3014, with the efficiency of albumin removal being 99.4%. Subsequently, new columns were developed to remove 12, 14, 20 and more proteins and IgY antibody was employed. In the past, columns were available to immunodeplete up to 81 of the most abundant proteins; however many of these products have been discontinued.

Fourteen of the most abundant proteins included in Seppro® IgY14 (Sigma–Aldrich, Inc.) are albumin, IgG, fibrinogen, transferrin, IgA, IgM, apolipoprotein A-I and II, haptoglobin, a1 antitrypsin, a1 acid-glycoprotein, a2 macroglobulin, apolipoprotein B, and complement C3. Specific removal of these 14 proteins depletes ~95% of the total protein mass from human serum, plasma or cerebrospinal fluid (CSF). Twenty-two of the most abundant proteins constitute ~99% of the total mass of proteins, thus, based on HPPP, leaving at least 2998 proteins in the remaining 1% [3]. The Seppro SuperMix System (Sigma–Aldrich, Inc.) was developed by immunizing chickens with a flow-through fraction of IgY-12 or IgY14 column and constructing the column with affinity-purified IgY antibodies against the flow-through proteins of IgY-12 or IgY14. The goal of the SuperMix System is to remove medium-abundant proteins from plasma/serum/CSF samples that were already immunodepleted using IgY12 or IgY14 columns. It is important to note that because antibodies used to make the SuperMix System are not fully standardized, the immune response of the particular chicken being immunized may vary. The subsequent introduction of variability by sample preparation may mask differences resulting from factors such as treatment, disease development, etc.

An alternative technique was proposed by Kovac and coworkers [4], and is based on Blue Sepharose 6 Fast Flow affinity chromatography using an XK 26/40 column in the AKTA liquid chromatography system to immunodeplete albumin from 500 ml of human plasma. The authors reported that based on SDS-PAGE analysis, the majority of albumin was removed; however, a majority of the depleted albumin also contained albumin-associated proteins and proteins showing affinity to Blue Sepharose.

When deciding to use either LC columns or spin columns, there are several factors to consider. First the investigator needs to determine how many proteins need to be depleted for their experimental design; would removal of IgG and albumin be sufficient, or is it necessary to remove more abundant proteins? Second, the investigator needs to know how great a volume of plasma needs to be depleted. For example, if the plasma is from a mouse, there is a small volume and a spin column would be sufficient. LC columns can handle larger volumes for samples such as human plasma. Capacity is discussed in further detail in the next section. Another detail to consider is that LC columns do require an High Performance Liquid Chromatography (HPLC) system, whereas the spin columns need a standard centrifuge. Third, cost must also be taken into consideration as there is quite a range of prices for LC columns versus spin columns.

6.3 Capacity of Immunodepletion Columns and Other Devices

The reference range for total protein in human plasma is 60–85 g/l. The term "reference range" in this case is a value used to interpret medical tests in clinical biochemistry. As much as this broad range can be considered normal and used in clinical practice, it has significant consequences for constructing immunodepletion columns or other devices and optimizing their quality control. In most cases, capacity of immunodepletion columns is given by manufacturer in microliters of serum/plasma. For example, the capacity of The Seppro® IgY 14 Liquid Chromatography 5 (LC5) Column is 100 μl of normal human serum or plasma. We can only assume, based on an average range of 60–85 g/l, that this 100 μl equals 6.0–8.5 mg of protein. If a sample from any given patient has less protein due to ongoing disease, we will not overload the column; however, if a patient has a higher level of proteins, eg, in paraproteinaemia, Hodgkin's lymphoma or leukemia, less volume of sample needs to be used or we assume again that the tolerance of protein capacity of such a column is broad enough that we can load more than 8.5 mg of protein and immunodepletion will be complete based on efficiencies provided by the manufacturer. Nevertheless, we do not have a fast and easy method to test whether our sample was properly immunodepleted. The situation is even more complicated when we try to immunodeplete CSF. CSF contains 10 to 100 times less protein mass than serum/plasma, which is yet another broad range. Eventually we end up with a methodology of immunodepletion that is based on wide range concentrations of total protein. Therefore, the amount of microliters of serum/plasma/CSF to be loaded onto the immunodepletion column needs to be assessed conservatively, as protein concentration is not measured and only volume is used as a measure of quantity.

6.4 Reproducibility

Reproducibility of each step in a multistep proteomic profiling experiment is critical and is associated with variability of all parameters. However, we need to keep in mind that in proteomics studies there are two major sources of variability: technical and biological. The impact of technical variability has decreased in last decade due to development of standardized protocols (kits), robotics, in particular autosamplers, as well as overall improvement of quality of instrumentation, supplies and reagents. In chromatographic resin (packing) used to make columns or devices for immunodepletion, the antibodies are oriented on the surface of solid beads and chemically cross-linked via the Fc region, and as a result the Fab interacting regions are exposed; covalent cross-linking also prevents leaching. These resins need to be used with great care and stored with sodium azide to protect from microbial growth if the column is not used daily. Based on our experience [5–9], washing column with sodium azide every 15 to 20 cycles is good practice and extends the life of a column.

While technical variation can be minimized, there is little that can really be done to reduce biological variability, which increases with increased complexity of organisms and/or biological systems under investigation. One way to offset biological variability in biomedical research is to use transformed cell lines, established viral or bacterial strains and in-bred animals. This is based on the assumption that if two laboratories use the same bacterial strain grown under the same conditions, variability from this part of experiment is minimized and the results of manipulating such systems can be meaningfully compared. This, however, does not apply to humans and nonhuman primates. The assembly of human subject cohorts participating in clinical studies can be only based on a set of predetermined clinical evaluations, which usually do not cover all variability. Furthermore, the use of predetermined conditions for collection of biological samples (eg, blood collected at same time of day after 12 h of fasting) cannot result in the same concentration of highly abundant proteins and percentage of albumin. If these subjects return for the second visit, the variable levels of highly abundant proteins present in the biological sample will affect the amount of proteins removed by immunodepletion during sample processing. This needs to be considered while planning and executing proteomic experiments, in particular those using body fluids.

SDS-PAGE separation followed by any type of protein staining might be a good visual measure; however, the pure analytical value of gel-based densitometry has many limitations, including a low level of precision. This fundamental analytical aspect of immunodepletion will have a profound effect on any type of downstream quantitation.

6.5 Quality Control of Immunodepletion

Besides comparison of LC and 1 Dimensional Electrophoresis (1DE) profiles there is no suitable method to monitor the quality of immunodepletion. 1DE can indicate whether there is any residual of depleted proteins present in the flow-through fraction if the gel-staining method is sensitive enough. For quality control, we excise a band from 1DE gel

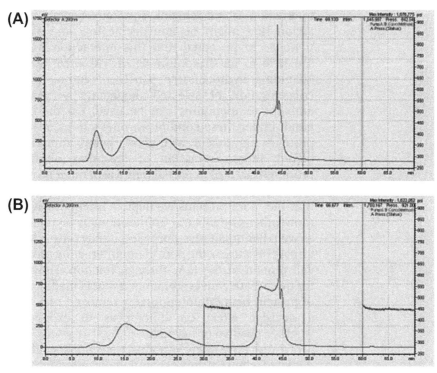

Figure 6.1 Reproducibility of immunodepletion of plasma samples using a Seppro® IgY14 column. (A) Representative profile of immunodepletion of one of the first 1–10 plasma samples. (B) Representative profile of immunodepletion of one of last 90–100 plasma samples.

where we would expect albumin (the most abundant protein) to be, perform in-gel digestion and analyze such sample using nano-LC-MS/MS. We have found this approach sensitive enough to show a slowly increasing number of albumin peptides after immunodepletion of 120 plasma samples using a column that was recommended by the manufacturer to perform according to specification to up to 100 samples. Comparison of LC profiles as shown in Fig. 6.1 is a crude assessment and can indicate only loss of more than 50% of column performance.

6.6 Albuminome

Immunodepletion is based on interaction of proteinaceous antigen with immunoglobulin, which is a protein as well. Although it is a highly specific interaction, it still remains a protein–protein interaction that might be nonspecific to some extent. It needs to be noted that this interaction will be sensitive to harsh conditions of 8 M urea, SDS or guanidine hydrochloride (Gu-HCl), and a limited concentration of selected detergents as well as ambient temperatures can be used. On the other hand, under these conditions, interactions of many proteins may occur with those being immunodepleted; thus elimination of the most abundant proteins may lead to partial removal of other proteins affecting quantification. Therefore, we can conclude that the more proteins we deplete intentionally, the more other proteins will unintentionally be removed as well. For qualitative purposes, it may have a lower impact because some pool of noninteracting proteins will remain in the flow-through fraction. However, for quantitative measurements, even a small portion of protein being unintentionally removed may have a decisive effect on differences in expression, whether positive or negative.

To test how many proteins were nonspecifically depleted when we applied human plasma samples to an IgY14 column, we took 50 μg of protein from an eluted fraction, fragmented by trypsin digestion, fractionated the resulting peptides using 24 well OFFGEL (based on pI) and each fraction was further analyzed using RP-nano-LC MS/MS. Spectra were searched against UniRef90 database with Proteome Discoverer (Sequest algorithm). We identified 96 proteins represented by at least one unique medium and one unique high-confidence peptide (Table 6.1). Because we used plasma from HIV-1 infected individuals, we identified such proteins as gp160, gp120 and Pol. Interestingly, many of proteins listed in this table are putative and their records in UniProt/TrEMBL still have the status "unreviewed."

Because albumin is known for its interactions with many molecules, including proteins, and is the most abundant protein in plasma, it is correctly assumed that the eluted fraction containing the most abundant proteins also contains other co-immunodepleted proteins. This fraction is called "albuminome," although it also refers to proteins co-immunodepleted due to interactions with other

Table 6.1 Human plasma proteins co-immunodepleted with an IgY14 column [10]

Accession	Mw (kDa)	Calculated pI	Protein name
A0N5G5	12.8	8.97	Rheumatoid factor D5 light chain (fragment)
A4D1A8	410.9	5.40	Similar to piccolo protein (aczonin)
A6NCL1	37.9	6.20	Geminin coiled-coil domain-containing protein 1
A6NKB1	3711.3	6.52	Titin
A6YID4	57.0	8.63	Fibronectin fragments or splice variant C (fragment)
A8K008	51.6	8.16	cDNA FLJ78387
B4DIE5	83.8	9.13	cDNA FLJ60561, highly similar to complement C4-B
B5A928	24.5	7.80	Soluble VEGFR3 variant 3
B5ANL2	94.9	8.43	Envelope glycoprotein gp160, gp120, or (fragment)
B6RP19	10.7	7.12	Pol
C9JEX1	43.8	6.43	Lysyl-bradykinin
C9JLB1	15.6	4.70	Uncharacterized protein
C9JU00	14.0	7.20	Uncharacterized protein
D6W5M4	78.6	8.29	Jumonji domain containing 1A, isoform CRA_b
E0VB03	188.4	9.41	Putative uncharacterized protein
E0VCF0	63.9	7.24	TFIIH basal transcription factor complex subunit, putative
E0VEP4	175.4	5.86	Dynamin-associated protein, putative
E0VG64	69.6	6.44	Paramyosin, putative
E0VHH7	150.9	8.25	Putative uncharacterized protein
E0VMF5	163.7	7.33	Putative uncharacterized protein
E0VMQ5	123.8	8.10	FYVE-containing protein, putative
E0VNC2	373.3	5.39	Putative uncharacterized protein
E0VPE6	172.6	6.38	Putative uncharacterized protein
E0VQH5	47.2	5.72	Putative uncharacterized protein
E0VVV4	67.6	9.32	Putative uncharacterized protein
E0VVX3	510.2	5.44	Low-density-lipoprotein receptor, putative

Continued

Table 6.1 *(continued)*

Accession	Mw (kDa)	Calculated pI	Protein name
E0VZE1	68.8	8.92	Putative uncharacterized protein
E0W0B2	156.1	7.87	DNA repair protein RAD50, putative
E0W0Z3	261.7	7.28	Fatty acid synthase, putative
E0W1F4	206.5	7.44	Putative uncharacterized protein
E7EMG0	211.2	5.03	Protocadherin-15
E9PHV3	77.7	9.11	DC-STAMP domain-containing protein 1
F5H7R6	117.7	8.62	Bloom syndrome protein
O00444	108.9	8.62	Serine/threonine-protein kinase PLK4
O14894	20.8	8.13	Transmembrane 4 L6 family member 5
O60229	340.0	6.07	Kalirin
O60673	352.6	8.47	DNA polymerase zeta catalytic subunit
O60765	69.2	9.57	Zinc finger protein 354A
P00742	54.7	5.94	Coagulation factor X
P01008	52.6	6.71	Antithrombin-III
P02747	25.8	8.41	Complement C1q subcomponent subunit C
P02787	77.0	7.12	Serotransferrin
P02790	51.6	7.02	Hemopexin
P04003	67.0	7.30	C4b-binding protein alpha chain
P04150	85.6	6.38	Glucocorticoid receptor
P04196	59.5	7.50	Histidine-rich glycoprotein
P05090	21.3	5.15	Apolipoprotein D
P05155	55.1	6.55	Plasma protease C1 inhibitor
P06725	62.9	7.24	65 kDa phosphoprotein
P06727	45.4	5.38	Apolipoprotein A-IV
P08603	139.0	6.61	Complement factor H
P10909	52.5	6.27	Clusterin
P16885	147.8	6.64	1-phosphatidylinositol-4,5-bisphosphate phosphodiesterase gamma-2
P26676	256.2	6.98	RNA-directed RNA polymerase L
P27169	39.7	5.22	Serum paraoxonase/arylesterase 1
P35579	226.4	5.60	Myosin-9
P68871	16.0	7.28	Hemoglobin subunit beta
Q01954	110.9	7.36	Zinc finger protein basonuclin-1
Q13064	55.6	5.73	Probable E3 ubiquitin-protein ligase makorin-3

Table 6.1 *(continued)*

Accession	Mw (kDa)	Calculated pl	Protein name
Q13349	126.7	5.77	Integrin alpha-D
Q14683	143.1	7.64	Structural maintenance of chromosomes protein 1A
Q15149	531.5	5.96	Plectin
Q15911	404.2	6.20	Zinc finger homeobox protein 3
Q3KRA7	28.8	9.44	FGA protein (fragment)
Q3L8U1	325.8	7.01	Chromodomain-helicase-DNA-binding protein 9
Q4R6T2	47.8	8.43	Testis cDNA, clone: QtsA-17169, similar to human complement component 1
Q59EK0	57.9	8.24	Epsilon isoform of regulatory subunit B56, protein phosphatase 2A variant (fragment)
Q5BJF6	95.3	7.62	Outer dense fiber protein 2
Q5JYW1	28.6	7.74	Forkhead-associated (FHA) phosphopeptide binding domain 1
Q5T4S7	573.5	6.04	E3 ubiquitin-protein ligase UBR4
Q5T8M7	37.8	5.58	Actin, alpha 1, skeletal muscle
Q5VWQ8	131.5	8.72	Disabled homolog 2-interacting protein
Q68CN4	51.5	7.56	Putative uncharacterized protein DKFZp686E23209
Q68CX6	235.3	5.74	Putative uncharacterized protein DKFZp686O13149
Q6GMX0	25.8	7.97	Putative uncharacterized protein
Q6GMX6	51.1	8.69	IGH@ protein
Q6LBZ1	19.9	8.15	MRNA for apolipoprotein E (apo E) (fragment)
Q6MZU6	51.1	7.71	Putative uncharacterized protein DKFZp686C15213
Q6N094	52.6	8.18	Putative uncharacterized protein DKFZp686O01196
Q6P5S8	25.8	6.33	IGK@ protein
Q6PGP7	175.4	7.53	Tetratricopeptide repeat protein 37
Q70EK8	120.7	7.59	Inactive ubiquitin carboxyl-terminal hydrolase 53
Q7Z7A1	268.7	5.55	Centriolin
Q8BB47	165.2	9.80	Immediate-early protein 2
Q8IZK6	65.9	7.61	Mucolipin-2

Continued

Table 6.1 *(continued)*

Accession	Mw (kDa)	Calculated pI	Protein name
Q8NDH2	277.8	9.31	Coiled-coil domain-containing protein 168
Q8NHA5	96.9	7.96	Seven transmembrane helix receptor
Q8WXI7	2351.2	6.00	Mucin-16
Q8WY24	53.6	6.67	SNC66 protein
Q92878	153.8	6.89	DNA repair protein RAD50
Q9BQ02	50.9	5.02	NCL protein
Q9HAR2	161.7	6.44	Latrophilin-3
Q9NQP4	15.3	4.53	Prefoldin subunit 4
Q9NYP9	25.8	5.20	Protein Mis18-alpha
Q9P1H1	17.1	8.82	DnaJ (Hsp40) homolog, subfamily A, member 4, isoform CRA_a
Q9UFH2	511.5	5.77	Dynein heavy-chain 17, axonemal

abundant proteins. Therefore, albuminome was a subject of several systematic studies that provided us with some insights into the composition of co-removed subproteomes [4,11,12]. Our considerations here are more focused on quantitative than qualitative effects of immunodepletion and other methods. From our laboratory practice, we conclude that the capacity and performance of IgY-based columns are much higher than recommended by the manufacturers. This is because of the large margin of specification resulting from lack of adequate analytical measures, relative fragility of antibodies and susceptibility to various types of damages. Although we did not, nor do we suggest to, load more sample than recommended be the manufacturers, we used the column for more cycles (injections). We started observing slow deterioration of column capacity after 120 cycles, while 100 were guaranteed by the manufacturer. This constitutes 20% of tolerance. Importantly, we always paid attention to plasma/serum/CSF sample preparation such as thorough delipidation and clarification using 0.22-μm pore spin filters. We used 1DE to monitor efficiency of albumin removal; however, we also constantly monitored how many unique albumin peptides we identified in immunodepleted samples and used that as indicator of column performance.

6.7 Summary

When working with serum/plasma/CSF or other serum-like fluids, eg, synovial fluid, a reduction of a high range of protein concentration is required and immunodepletion seems to be the current method of choice until new methods are developed. Therefore when designing proteomic experiments, we must acknowledge the analytical specificity or drawbacks of this component of sample processing. Factors such as the broad range of concentration of proteins under "normal physiological" conditions, relatively broad range of quantitative tolerance of the method itself, and the lack of quick and precise measures of antibody-based column performance, may detract or enhance quantitative variability proteins. On the other hand, both experimental and control samples are immunodepleted under the same conditions, and thus co-depletion should lessened affect in identification of potential biomarkers.

References

[1] Omenn GS, States DJ, Adamski M, Blackwell TW, Menon R, Hermjakob H, et al. Overview of the HUPO Plasma Proteome Project: results from the pilot phase with 35 collaborating laboratories and multiple analytical groups, generating a core dataset of 3020 proteins and a publicly-available database. Proteomics 2005;5(13):3226–45.

[2] Anderson NL, Anderson NG. The human plasma proteome: history, character, and diagnostic prospects. Mol Cell Proteomics 2002;1(11):845–67.

[3] Borg J, Campos A, Diema C, Omenaca N, de Oliveira E, Guinovart J, et al. Spectral counting assessment of protein dynamic range in cerebrospinal fluid following depletion with plasma-designed immunoaffinity columns. Clin Proteomics 2011;8(1):6.

[4] Kovacs A, Sperling E, Lazar J, Balogh A, Kadas J, Szekrenyes A, et al. Fractionation of the human plasma proteome for monoclonal antibody proteomics-based biomarker discovery. Electrophoresis 2011;32(15):1916–25.

[5] Wiederin JL, Donahoe RM, Anderson JR, Yu F, Fox HS, Gendelman HE, et al. Plasma proteomic analysis of simian immunodeficiency virus infection of rhesus macaques. J Proteome Res 2010;9(9):4721–31. Epub 2010/08/04.

[6] Wiederin JL, Yu F, Donahoe RM, Fox HS, Ciborowski P, Gendelman HE. Changes in the plasma proteome follows chronic opiate administration in simian immunodeficiency virus infected rhesus macaques. Drug Alcohol Dependence 2012;120(1–3):105–12. Epub 2011/08/09.

[7] Schlautman JD, Rozek W, Stetler R, Mosley RL, Gendelman HE, Ciborowski P. Multidimensional protein fractionation using ProteomeLab PF 2D for profiling amyotrophic lateral sclerosis immunity: a preliminary report. Proteome Sci 2008;6:26. Epub 2008/09/16.

[8] Rozek W, Horning J, Anderson J, Ciborowski P. Sera proteomic biomarker profiling in HIV-1 infected subjects with cognitive impairment. Proteomics Clin Appl 2008;2(10–11):1498–507. Epub 2008/10/01.

[9] Rozek W, Ricardo-Dukelow M, Holloway S, Gendelman HE, Wojna V, Melendez LM, et al. Cerebrospinal fluid proteomic profiling of HIV-1-infected patients with cognitive impairment. J Proteome Res 2007;6(11):4189–99. Epub 2007/10/13.

[10] Pottiez G, Jagadish T, Yu F, Letendre S, Ellis R, Duarte NA, et al. Plasma proteomic profiling in HIV-1 infected methamphetamine abusers. PLoS One 2012;7(2):e31031. Epub 2012/02/24.

[11] Liu B, Qiu FH, Voss C, Xu Y, Zhao MZ, Wu YX, et al. Evaluation of three high abundance protein depletion kits for umbilical cord serum proteomics. Proteome Sci 2011;9(1):24. Epub 2011/05/11.

[12] Gundry RL, Fu Q, Jelinek CA, Van Eyk JE, Cotter RJ. Investigation of an albumin-enriched fraction of human serum and its albuminome. Proteomics Clin Appl 2007;1(1):73–88. Epub 2007/01/01.

7

GEL ELECTROPHORESIS

A. Drabik and A. Bodzoń-Kułakowska
AGH University of Science and Technology, Krakow, Poland

J. Silberring
AGH University of Science and Technology, Krakow, Poland;
Polish Academy of Sciences, Zabrze, Poland

CHAPTER OUTLINE

7.1 FUNDAMENTALS OF GEL ELECTROPHORESIS   117
7.1.1 Introduction   117
7.1.2 Electrophoresis Conditions   119
7.1.3 Agarose Gel Electrophoresis   119
7.1.4 Sample Preparation   120
7.1.5 Separation Conditions   120
7.1.6 Native Polyacrylamide Gel Electrophoresis   121
7.1.7 Electrophoresis in Denaturing Conditions   123
7.1.8 Sample Preparation Prior to SDS-PAGE   124
7.1.9 Staining Techniques   124
7.1.10 Fluorescent Staining   126
7.1.11 Isotope Labeling   127
7.1.12 Data Storage   127
References   127

7.2 TWO-DIMENSIONAL GEL ELECTROPHORESIS   128
7.2.1 Introduction   128
7.2.2 First Dimension of Two-Dimensional
      Electrophoresis: The Isoelectric Point   129
7.2.3 Second Dimension of Two-Dimensional
      Electrophoresis: Molecular Weight   131
7.2.4 Gel Staining   133
7.2.5 Pros and Cons of Two-Dimensional Gel
      Electrophoresis   135
7.2.6 Quantitation of Protein Using Two-Dimensional
      Gels   136

Proteomic Profiling and Analytical Chemistry. http://dx.doi.org/10.1016/B978-0-444-63688-1.00007-0

7.2.7 Difference Gel Electrophoresis 139
**7.2.8 Fluorescent Dyes Used in Difference Gel
Electrophoresis 140**
7.2.9 Internal Standard 141
**7.2.10 Pros and Cons of Difference Gel
Electrophoresis 142**
References 142

FUNDAMENTALS OF GEL ELECTROPHORESIS

A. Drabik
AGH University of Science and Technology, Krakow, Poland

J. Silberring
AGH University of Science and Technology, Krakow, Poland;
Polish Academy of Sciences, Zabrze, Poland

7.1.1 Introduction

Electrophoretic separation is based on the migration of unbalanced charged molecules in an electric field and is the most frequently used dispensation method in the study of proteins and nucleic acids. It is widely used in biochemistry, molecular biology, pharmacology, criminal medicine, diagnostics and food quality control. Electrophoresis can be used for macromolecule isolation in complex biological systems, as well as a tool for determining molecular weight (MW) and detecting structural and charge-state modifications. It can also be applied as a sample purity control, as well as a discovery tool for proteins, nucleic acids and large peptides.

The main premise of electrophoretic separation is application of an electric field that forces molecules to move through gel pores, separating them based on their MW and total particle charge. Large-molecular-weight molecules are slowed down on the basis of gel pore size (Tables 7.1.1 and 7.1.2); more specifically, larger-molecular-weight molecules are "trapped" in regions of the gel with a higher percent concentration [1]. The migration ratio is a constant value that is directly proportional to the electric current, shape

Table 7.1.1 Molecular separation range as a function of agarose gel concentration

Agarose concentration (g/100 ml)	Molecule size (kDa)
0.3	5–60
0.6	1–20
0.7	0.8–10
0.9	0.5–7
1.2	0.4–6
1.5	0.2–3
2	0.1–2

Table 7.1.2 Acrylamide concentration correlation with separated species molecular weight

$$\% \ C = \frac{\text{bisacrylamide}[g] \times 100}{\text{acrylamide}[g] + \text{bisacrylamide}[g]}$$ cross-linking agent mass concentration

%C	Mw (kDa)
7	50–500
10	20–300
12	10–200
15	3–100

and size of separated species, hydrophobicity, ionic strength, viscosity, and temperature, as defined here:

$$\mu = \frac{V}{E} = \frac{Z}{f}$$

$$f = 6\pi\eta r$$

where μ is electrophoretic mobility, V is migration speed, E is electric field strength, Z is total molecular charge, f is friction coefficient, η is viscosity, and r is molecule radius.

Application of a specially designed electrophoretic power supply enables the user to keep a constant value of a selected parameter: voltage, current, or power. During the separation process, electrolyte resistance is reduced by temperature increases, while a reduction in the number of ions and their arrangement order in the gel can increase the resistance. When the temperature is elevated, the separation time is extended, which can cause molecules to diffuse, thereby decreasing the resolving power.

7.1.2 Electrophoresis Conditions

Electrophoretic separation can be performed under various conditions, which can either maintain protein complexes intact, or dissociate them to obtain and identify particular components. The desired outcome is dependent on the gel types, detergents and cathode buffers; these variable parameters for different types of electrophoresis are described in more detail later in the chapter.

7.1.3 Agarose Gel Electrophoresis

Agarose gel electrophoresis is a method of choice for large molecule separation over 1 million Da. Acrylamide cannot be used for this purpose, because it remains liquid at the concentration required for the appropriate separation of high-molecular-weight analytes. The movement of molecules through an agarose gel is dependent on the size and charge of separated particles, as well as the pore size present in the gel. The observed migration is also affected by the type of electrophoresis buffer, especially its ionic strength. The electrical conductance of the gel is dependent on the presence of various ions, including those present in the sample. Gel polymerization is based on heating the agarose solution (Table 7.1.1) to a temperature higher than 40°C. Polysaccharide is solidified again after cooling to room temperature. Usually, the separation process is positioned horizontally (Fig. 7.1.1). Undoubtedly, major advantages of this particular technique are easy and rapid preparation of the gels and the possibility of

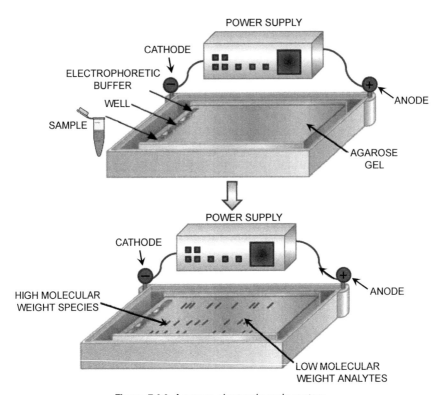

Figure 7.1.1 Agarose electrophoresis system.

high-molecular-weight species fractionation. The combination of agarose gel with polyacrylamide gel electrophoresis enabled creation of the genome maps and facilitated the Human Genome Project.

7.1.4 Sample Preparation

The protocol for sample preparation for agarose gel electrophoresis is straightforward and fast. Analytes are mixed 1:1 (v/v) with loading buffer, which consists of glycerol, bromophenol blue dye, and electrophoretic buffer. High salt content should be avoided as it could affect the separation process [2].

7.1.5 Separation Conditions

Power supply settings are typically 5 V/cm. In other words, if the electrodes are 10 cm apart, the alternating current should be set to 50 V. However, it is possible to adjust separation conditions to a particular sample requirement.

7.1.6 Native Polyacrylamide Gel Electrophoresis

This type of electrophoretic separation allows for the fractionation of species based on their surface charge density in a nondenaturing environment. The stabilizing forces of protein complexes are identical to those that play a role in protein folding (eg, ionic interactions, dipole interactions, hydrogen bonds, van der Waals forces, hydrophobic interactions, and water-mediated interactions between residues). Therefore, the use of anionic detergents such as sodium dodecyl sulfate (SDS), which disrupts such interactions, is not recommended for intact protein analyses. The ability to preserve mild detergent-stable protein complexes is the major benefit of this method. Proteins must have a negative charge for effective migrations; as a rule, the greater a negative charge is located on their surface, the more rapidly protein will travel. The method described can be applied to monitoring protein–protein interactions, charge changes, conformational alterations, formation of aggregates, and examination of the stability of the analyzed complexes [3]. The MW range of the complexes varies between 10 kDa and 10 MDa. Proteins might be recovered from the gel with the use of electroelution or passive diffusion.

All native electrophoresis systems commonly use acrylamide-based gradient gels for protein separation. A polyacrylamide gel (PAGE) is polymerized after combining two toxic compounds, acrylamide and bisacrylamide, and subsequent addition of the cross-linking agent ammonium persulfate, and N,N,N′,N′-tetramethylethylenediamine (TEMED), which serves as a catalyst. The separation is highly dependent on pore size, which can be adjusted by changing the acrylamide concentration (Table 7.1.2). The bisacrylamide concentration for the highest-density pores is 5%, and each modification (increase or decrease of bisacrylamide content) causes pore enlargement. Polymerization of the gel is achieved after 30 min at room temperature; however, complete cross-linking is obtained after 12 h. It is worth noting that atmospheric oxygen acts as an inhibitor of this process; hence it is necessary to protect the gel by covering its surface with water or butanol. Because the polymerization reaction is exothermic, the best results are obtained by keeping the system in a refrigerator. Typically, the separation process is performed in a vertical position (Fig. 7.1.2).

From a practical point of view, the only difference between the native electrophoresis techniques blue native electrophoresis (BNE), clear native electrophoresis (CNE), and high-resolution clear electrophoresis (hrCNE) is the type of cathode buffer used.

The most popular technique, BNE, was initially developed to isolate complexes from purified mitochondria. In the first step, protein complexes are exposed to mild detergent necessary for their solubilization [4]. The anionic dye Coomassie Brilliant Blue G-250 (CBB) is added to the cathode buffer. It adds the negative charge to the proteins and modifies

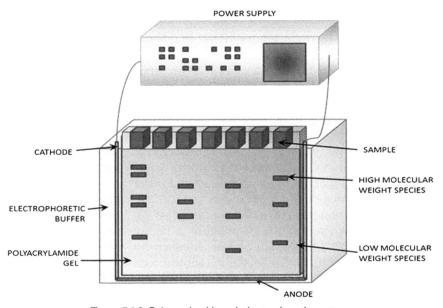

Figure 7.1.2 Polyacrylamide gel electrophoretic system.

their surface. CBB binds to basic amino acids by a combination of hydrophobic interactions and heteropolar bonding, forcing proteins to move toward the anode, independently of their intrinsic pI. Therefore even basic, water-soluble proteins such as cytochrome c (pI 10.7) can bind to CBB and move to the anode in running buffer pH 7.5.

Some proteins cannot be resolved with the use of BNE, such as high posttranslationally modified proteins, which do not bind to CBB—for example, mucins, and those having a neutral or basic pI. Those proteins show cathodic migration and are lost in the cathode buffer. The main disadvantage of CBB application is the formation of mixed anionic micelles in the presence of detergents (Triton X-100, dodecyl maltoside, digitonin) that do not bind to the molecules.

During CNE, separation occurs due to the intrinsic charge of a given protein; therefore no CBB is added to the sample and cathode buffer. Water-soluble proteins and complexes that do not require detergents for solubilization give rise to the high-resolution bands in CNE. One benefit of CNE is the capability of in-gel catalytic activity assays and analyses of fluorescent-labeled proteins. However, the main drawback of this system is the limitation of separating only acidic proteins with a pI value lower than 7. All basic proteins are lost during CNE separation.

In the case of hrCNE, a mixture of anionic and neutral detergent micelles is applied, as it is a charge shift technique like BNE. Still, not all water-soluble proteins bind to these anionic compounds, and as a result, there is no improvement of hrCNE compared to CNE or BNE.

7.1.7 Electrophoresis in Denaturing Conditions

Sodium dodecyl sulfate-polyacrylamide acrylamide electrophoresis (SDS-PAGE) is one of the most frequently used techniques of protein separation. It allows for further quantitative and qualitative identification with the use of Western blotting or mass spectrometry. The sample migrates in the presence of ionic detergent SDS, which is responsible for protein denaturation and destruction of complex interactions, as well as electrophoretic separation of particles in an independent manner to their charge. SDS provides a homogenous negative charge to all of the separated proteins.

7.1.8 Sample Preparation Prior to SDS-PAGE

SDS-PAGE is more laborious than agarose electrophoresis when it comes to both gel and sample preparation. Proteins must be reduced (disulfide bonds must be disrupted) prior to SDS-PAGE using common reducing agents such as dithiothreitol (DTT), beta-mercaptoethanol (BME), dithioerythritol (DTE), and tris 2-carboxyethyl phosphine (TCEP). Iodoacetamide is added to block free thiol groups.

A high salt concentration is undesirable in any kind of electrophoretic sample because it raises the current during the separation process. As a result, smears and heterogeneous bands are observed, as well as protein precipitation. The proper sample concentration should be applied for high-resolution and good-quality results. Moreover, the presence of abundant proteins can interfere with sample constituents that can be found at minor concentrations. The amount of each component cannot be higher than 10 μg. Supplementation with glycerol, along with bromophenol blue dye, makes sample application more convenient. Glycerol acts as a weight for the applied sample; the dye allows for visualization as well as pH monitoring. In summary, sample preparation for SDS-PAGE is mainly based on protein solubilization and denaturation.

7.1.9 Staining Techniques

Staining procedures are part of the electrophoretic process since proteins or nucleic acids are not detected in visible light. The most important properties of protein visualization methods are high sensitivity, low detection limit, high dynamic range for quantitation accuracy, reproducibility and compatibility with postelectrophoretic protein identification systems (Tables 7.1.3 and 7.1.4).

Table 7.1.3 Advantages and disadvantages of electrophoresis

Advantages	Disadvantages
Simultaneous qualitative/quantitative analysis	Time-consuming and laborious
Compatibility with MS and Western blotting techniques	Lack of automation
	High toxicity of acrylamide and its influence on MS analysis

Table 7.1.4 Main obstacles during electrophoretic separation

Problem	Explanation
Incomplete acrylamide gel polymerization	Wrong proportions among gel components, expired ammonium persulfate (APS) activity
Too-dense gel causes disruption of wells during comb removal	High acrylamide concentration
Gel presence inside wells	Incompletely polymerized gel or incorrect comb size
Low quality of separation	Insufficient time of separation, wrong gel concentration, incomplete well filling
Faint staining	Decreased dye concentration, too short a time of staining, wrong type of applied dye, presence of SDS on the surface of the gel that prevents staining
Irregular staining	Insufficient mixing during staining or dye solution filtering
Bands smearing	Precipitation of proteins, addition of SDS and reducing agent are required, presence of air bubbles
Curve bands shape	Uneven gel formation, too high temperature, too low or too high APS concentration

Coomassie Brilliant Blue is an organic dye able to detect proteins with lower detection levels of 8–10 ng. There are two types of CBB: G-250, which is characterized by a greenish color; and reddish R-250, which is less sensitive. A colloidal solution of CBB provides enhanced reproducibility and a destaining procedure is no longer required, as in case of an acidic CBB solution [5].

Silver staining is widely used in proteomic approaches, despite being a multistep, laborious process; however, its detection limit is higher than 100 pg in a single band. It is worth noting that the linear dynamic range of this dye is relatively low; for quantitative analysis, this staining technique may not be effective in covering a broader dynamic range of separated molecules.

7.1.10 Fluorescent Staining

SYPRO and Pro-Q dyes are popular among proteomic researchers because fluorescent stains show a remarkably wide dynamic range, to the extent of even four orders of magnitude, and therefore are useful for quantitative studies. Red and Orange SYPRO bind to the detergent molecules surrounding proteins, while SYPRO Ruby interacts with basic amino acids. The visualization process is very simple and rapid, involving only a single step. The limit of protein detection is 0.5–5 ng. Yet another benefit of fluorescent dye application is the possibility of observing posttranslational modifications. Pro-Q-Emerald stain allows for revealing glycosylation sites, and Pro-Q-Diamond is used in the investigation of phosphorylations.

Samples stained with fluorescent dyes produce characteristic differential patterns in difference gel electrophoresis (DIGE). Each of the stains represents one sample. This technique provides simultaneous comparative analysis of studied samples; for example, proteomes before and after treatment, or cancer patients versus noncancer patients, etc. DIGE eliminates run-to-run error in image evaluation.

7.1.11 Isotope Labeling

Radioactive isotope labeling ^{125}I, ^{123}I, ^{14}C, ^{35}S, ^{32}P, and ^1H takes place before electrophoretic separation. Because this approach is characterized by poor reproducibility and specific restrictions concerning radioactivity, it is not often applied to proteomic studies.

7.1.12 Data Storage

The best and simplest method for data documentation is as a scan or photo files. Polyacrylamide gels can be stored in 0.05% sodium azide solution ahead of lamination. Agarose gels are not suitable for long-term storage due to very fast dehydration, even after hermetic sealing, and such dry gel breaks easily.

In summary, electrophoretic separation can be performed for high-molecular-weight peptides, proteins and nucleic acids. Despite the problem with its automation, it can be applied to both quantitative and qualitative analyses of biological samples. Typically, the proteomic approach utilizes electrophoresis for further Western blotting or mass spectrometry. The fact that this separation technique has been widely applied for over 100 years is sufficient proof of its usefulness.

References

[1] Drabik A, Laidler P. One-dimensional gel electrophoresis. In: Kraj A, Silberring J, editors. Proteomics: Introduction to methods and applications. Hoboken (NJ, USA): John Wiley & Sons, Inc.; 2008.

[2] Bodzoń-Kułakowska A, Bierczyńska-Krzysik A, Dyląg T, Drabik A, Suder P, Noga M, et al. Methods for samples preparation in proteomic research. J Chromatogr B 2007;849:1−31.

[3] Wittig I, Schagger H. Native electrophoretic techniques to identify protein-protein interactions. Proteomics 2009;9:5214−23.

[4] Wittig I, Schagger H. Features and applications of blue-native and clear-native electrophoresis. Proteomics 2008;8:3974−90.

[5] Westermeier R, Marouga R. Protein detection methods in proteomic research. Biosci Rep 2005;25:19−32.

7.2

TWO-DIMENSIONAL GEL ELECTROPHORESIS

A. Bodzoń-Kułakowska
AGH University of Science and Technology, Krakow, Poland

J. Silberring
AGH University of Science and Technology, Krakow, Poland;
Polish Academy of Sciences, Zabrze, Poland

7.2.1 Introduction

The most common analytical techniques to separate intact proteins are one-dimensional electrophoresis (1DE) and two-dimensional electrophoresis (2DE) in polyacrylamide gel (PAGE). These techniques are performed under denaturating conditions using heat in the presence of sodium dodecyl sulfate (SDS) and reducing agents to break disulfide bonds. Proteins are separated according to their MW when they are unfolded by denaturation and acquire a negative charge from SDS. The general downside of gel electrophoresis is that after separation, proteins are trapped in the gel and typically need to be extracted for further analysis. Although other methods, such as size exclusion or reverse-phase high-performance liquid chromatography (SE-HPLC or RP-HPLC, respectively), can separate proteins, the resolution power of highly complex protein samples is lowered, thus making gel electrophoresis a more desirable method. A general downside of gel electrophoresis is that after separation, proteins are trapped in gel and their recovery is a challenge. But for proteomics experiments employing in-gel enzymatic fragmentation as a subsequent step, this is of lesser concern. Nevertheless, we have to keep in mind that extraction of peptide fragments from polyacrylamide gel is also associated with sample loss. Recovery from silver-stained bands or spots is usually lower than from samples stained with other protein-detecting dyes such as Coomassie BB or fluorescent dyes such as Sypro Ruby. In 2DE, two modes of separation are utilized: separation based on isoelectric points using immobilized pH gradient (IPG) strips, which is followed by SDS-PAGE gel electrophoresis [1–5].

7.2.2 First Dimension of Two-Dimensional Electrophoresis: The Isoelectric Point

Isoelectric focusing (IEF) separation is based on the simple mechanism that for any individual protein or peptide there is a pH point at which the molecule will have a net charge of zero and therefore will not migrate further in an electric field. In the basic region of the pH gradient, the acidic side chains of amino acid residues will show negative charges. In the acidic region, the basic side chains of amino acid residues will have positive charges. Mixtures of proteins to be separated are first denatured and then loaded onto the IPG strip, which has a linear pH gradient. IPG strips are polyacrylamide-based, and upon rehydration, proteins are driven into the strip along with the buffer. Application of electric current forces charged molecules to move toward the electrode of an opposite charge. During migration, protein/peptide side chains will gain or lose the net charge at their specific pH. A schematic representation of this process is shown in Fig. 7.2.1. An important factor influencing separation is the degree of protein unfolding. SDS cannot be used in IEF because it provides the same negative charge to all proteins; therefore denaturation is performed in the presence of 8 M urea. Theoretically, all proteins should be fully denatured in such conditions;

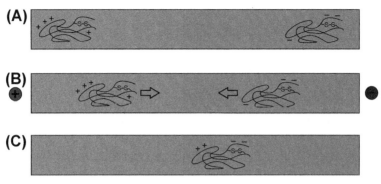

Isoelectric point

Figure 7.2.1 Isoelectric focusing. (A) The same protein placed at various places on the IPG strip, and the charge it obtains red (light gray in print version), low pH; blue (dark gray in print version), high pH. (B) Protein movement due to the electric current application. (C) Protein in its isoelectric point.

however, in real life one protein is represented by a population of molecules with partial denaturation. This leads to formation of residues that are "hidden" inside the protein structure and will not contribute to the overall net charge. This is observed as a series of horizontally located spots (train spots) quite often falsely interpreted as spots containing post-translational modifications (PTMs), which also create a similar separation pattern.

IPG strips are commercially available but restricted to the dimensions of the specified apparatus (same manufacturer). The gradient is formed on the thick gel strips by covalent incorporation of the gradient of buffering acrylamide derivatives into the polyacrylamide gel. This ensures stability of the pH gradient during electrophoresis. The pH range immobilized in the strips varies and depends on the purpose of separation. IPG strips can have a variety of ranges, from very broad (eg, pH 2.5–12) to very narrow (eg, pH 4–5). The desired pH depends on the chemistry of the proteins to be separated. For instance, acidic proteins may be separated on the pH 4–5 strips, whereas basic proteins can be focused between pH 6 and 11.

7.2.3 Second Dimension of Two-Dimensional Electrophoresis: Molecular Weight

The second dimension of 2DE separates proteins according to their MW. This separation takes place in a polyacrylamide gel matrix, which acts as a molecular sieve. The appropriate mesh in this sieve is made of the polymerized acrylamide, cross-linked by bisacrylamide. The size of pores in the polyacrylamide gel depends on the percentage amounts of acrylamide and bisacrylamide. A lower percentage of acrylamide (larger pore size) allows for separation of high-molecular-weight proteins. Higher acrylamide content (smaller pore size) is more suitable for resolving smaller proteins.

During the separation process it is possible to use nongradient gels and achieve excellent resolution for proteins in a narrow MW range. However, for separation of a complex mixture of proteins with a wide range of MW, it is advisable to use gradient gels. In gradient gels, the pore sizes decrease along the gradient and make it possible to achieve sharp protein separation for both large and small proteins (ie, 0.5–300 kDa) on the same gel. The separation process in the second dimension is based on protein migration caused by application of electric current through a polyacrylamide gel. During this process, large proteins with high MW are not able to migrate through the polyacrylamide sieve for a long distance and stay at the cathode end of the gel. Smaller proteins, with lower MW, migrate easily through the pores and travel toward the anode end of the gel (Fig. 7.2.2).

It is important to remember that proteins may have various shapes and they can migrate through the gel with various speeds, despite their MW. For example, a protein with an elongated shape may stay at the cathode end of the gel, even though it is not so "heavy." Alternatively, a protein might have a relatively high MW, but its structure may be compressed and thus permit travel faster through the gel.

Denaturation of proteins may minimize this "shape effect." Denaturation occurs with the reduction and

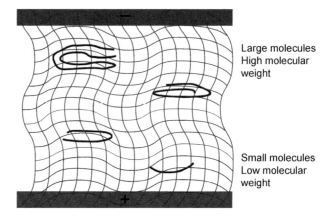

Figure 7.2.2 Proteins of different masses traveling through poly-acrylamide gel. −, Cathode; +, anode.

alkylation of disulfide bridges (Fig. 7.2.3), and this occurs after protein separation in the first dimension. The IPG strips with separated proteins are immersed in the solution of reducing agent, usually dithiothreitol (DTT), which reduces disulfide bridges responsible for the tertiary structure of proteins. To protect those residues from oxidation and formation of high-molecular-mass aggregates, iodoacetamide is added to binds to the free−SH moieties. Sodium dodecyl sulfate (SDS) is added to the DTT and iodoacetamide buffers, and it acts as an anionic surfactant by disrupting hydrogen bonds and blocking hydrophobic interactions to ensure protein unfolding. Thus, after reduction and alkylation in buffers with SDS, all proteins gain a similar, rod-like structure.

The question is how to force proteins with the net charge equal to zero following IEF of first-dimension separation and move them through the poly-acrylamide sieve. SDS is a molecule that possesses

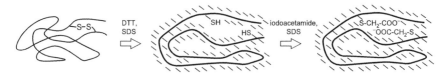

Figure 7.2.3 Reduction and alkylation of proteins. Blue (gray in print versions) rods represent SDS molecules, which are responsible for an overall negative charge and protein unfolding.

a negative charge at the end of its hydrophobic dodecyl group. And when SDS is added to the reduction and alkylation buffers, its hydrophobic chain interacts with nonpolar amino acids and the entire protein gains a strong negative charge, which masks its intrinsic charge. Moreover, SDS binds to the protein in proportion to its size, meaning that biomolecules with various masses gain the same charge density at their surface and may move in the electric field with the same mobility. This guarantees separation almost exclusively due to the protein molecular mass, regardless of its pI or shape.

There are several obstacles critical for the success of reproducible separations, the key being proper attachment of the IPG strip with separated proteins (first dimension) to the polyacrylamide gel (second dimension). Accurate attachment of these two elements is very important. The IPG strip, after reduction, alkylation and equilibration with SDS buffer, is applied on the top of polyacrylamide gel and fixed with molten agarose. This provides a good contact between polyacrylamide gel and the IPG strip. Then, an electric field is applied and proteins according to their net negative charge migrate from the cathode (negative) toward the anode (positive). This process ensures separation in the second dimension according to protein size and MW.

7.2.4 Gel Staining

Proteins do not absorb any wavelength from the visible range of the light spectrum and therefore the polyacrylamide gel with separated proteins is still colorless. Gel staining enables visualization of the proteins in a quantitative manner. An ideal dye should bind to the proteins noncovalently and in proportion to their concentration. It should possess a wide dynamic range and be sensitive enough to visualize low-abundant proteins. Additionally, it should not demonstrate saturation effects with highly abundant molecules, which can make normalization for quantitation difficult. An ideal dye should be compatible with the next steps of analysis (mass spectrometry in this case). There is no perfect staining method for 2D gel electrophoresis. For proteomics purposes, the most popular are still Coomassie Brilliant Blue and silver or fluorescent dyes, such as SYPRO Ruby (Table 7.2.1) [5].

Coomassie Brilliant Blue is able to detect 8–10 ng of protein and its dynamic range covers two orders of magnitude. It is preferred when relative amounts of protein are to be determined, because it binds stoichiometrically to proteins. Silver is the most sensitive approach to protein visualization and it is able to stain 0.1 ng of protein, but its dynamic range is narrower than two orders of magnitude and it is the least reproducible of all stains. Silver staining can also cause problems with downstream mass spectrometry analysis, since during the staining process glutaraldehyde is used during gel fixation and this substance can cross-link the proteins, decreasing the efficiency of trypsin digestion.

Fluorescence methods provide an alternative to Coomassie and silver staining. Since the measurement of light emission is much more sensitive than absorbance, fluorescence techniques are more sensitive than standard colorimetric techniques. The entire procedure is much simpler than silver staining, and the recovery of peptides for mass spectrometry is also higher. Sypro Ruby is one of the most frequently used fluorescent dyes. It can detect approximately 1 ng of protein per spot and with linearity spanning three orders of magnitude [6]. The problem with this

Table 7.2.1 Types of dyes used in staining of proteins separated on polyacrylamide gels

Staining	Detection limit	Linear response
Colloidal Coomassie blue	8–10 ng	More than 10^2
Silver	0.1 ng	Less than 10^2
Fluorescence	1 ng	Over 10^3

type of staining is that it demands special equipment for protein visualization, such as fluorescent scanners or ultraviolet light boxes, which increases the expense of this technique.

The more sensitive the staining technique, the less material required for visualization—which may translate into an insufficient amount of protein for identification using MS analysis. Such a small sample is also more susceptible to contamination and loss due to adsorption to glass or other types of test tubes.

7.2.5 Pros and Cons of Two-Dimensional Gel Electrophoresis

The main advantage of 2D gel electrophoresis is its ability to simultaneously visualize thousands of protein spots (from 500 to 3000 proteins during one analysis) originated from a sample. This technique provides a large amount of information since it is possible to determine the approximate MW and isoelectric point of each protein on the gel. Additionally, proteins with posttranslational modifications may be observed on the gel as horizontal or vertical spot clusters, because any modification may affect the MW and/or pI value. Quantitative measurements of protein expression between samples are possible after gel scanning and performing a comparative analysis of its image with appropriate software.

Apart from those advantages, there are several problems in protein separation using 2D gel electrophoresis. It is estimated that 70% of proteins identified on a 2D gel are located between 20 and 70 kDa [3]. Large proteins have difficulty entering the IPG strips with high efficiency, and therefore they cannot be detected during analysis. Basic proteins account for approximately one-third to one-half of all proteins in the cell, and the problem with their separation in 2D gel electrophoresis is that the commercially available basic IPG strips have decreased resolution compared to acidic strips [2]. Hydrophobic proteins, which constitute approximately 30% of cellular protein, may also add a challenge to 2D gel analysis since they are insoluble in standard protein-extraction solutions that are compatible with this system. It is estimated that only ~1% of integral membrane proteins could be separated in conventional 2D gel. It is worth noting that hydrophobic proteins make up to about 30% of the whole protein content of the cell [2]. Hydrophobic proteins are responsible for cell adhesion, metabolites and ion transport, and usually initialize signal transduction pathways by receiving the information from outside the cell. Because of their functions, they are often targets for new drugs, and a lack of their representation on the standard 2D gel may be a serious drawback [2]. These proteins may be separated using a different type of gel electrophoresis, such as 16-BAC/SDS-PAGE. Briefly, using discontinuous gel electrophoresis in an acidic buffer system, with the cationic detergent benzyldimethyl-n-hexadecylammonium chloride (16-BAC), allows for solubilization and separation of such proteins in the first dimension. After that, standard SDS/PAGE is applied for separation of proteins in the second dimension [7].

Using 2D gel electrophoresis, protein can co-migrate in the gel, caused by their similar physicochemical properties. This phenomenon may hamper protein identification and quantification and is frequently reported when using IPG strips with a wide range of pH values. Many major spots or streaks visible on this kind of gel appear to be separated in several spots on the gels with a wider pH range [8]. The resolution of the analysis may be improved by using a narrow pH gradient (so-called "zoom gels"), or applying the gel to a bigger area and more sensitive staining, or a combination of options [1].

7.2.6 Quantitation of Protein Using Two-Dimensional Gels

Software analysis of 2D gels is one of the main components of the proteomic approach and is a crucial step of the whole experiment. It allows for comparing two sets of gels from control and experimental samples. As a result, we gain information about certain sets of spots that significantly differ in their quantities between two samples and thus may serve as potential markers of the experimental condition. The focus of proteomics has shifted from being able to identify the maximum number of proteins to finding differences in protein expression and interactions that may lead to the discovery of potential biomarkers. Commercial software for identification of potential differences is usually a "black box" for the end-user, and there is minimal description given on the details of algorithms applied and minimal modification of parameters possible by the operator. Therefore, we will describe suggestions and provide a flowchart for efficient and statistically reliable procedures for identification of changes in protein profiles.

Software analysis consists of several steps (Fig. 7.2.4) [9,10]. First, the image of the gel must be acquired using scanners or other devices, such as charge-coupled devices, camera-based or laser-imaging instruments. Good-quality raw data is very important, as it impacts the final results. The next step of image processing is called image warping. Warping removes the variations in the same spot

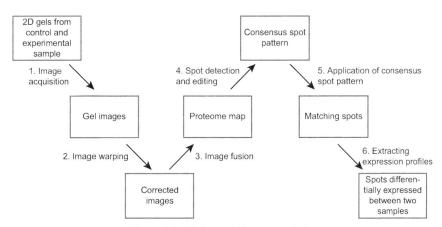

Figure 7.2.4 Scheme of image analysis.

positions on the gel replicates and is crucial for the correct spot matching. Similar regions or corresponding spots between gels are searched, vectors connecting corresponding points (spots) are determined and according to those vectors, the image is transformed. Artifacts like speckles or fingerprints may disturb this process, but modern software usually overcomes those problems.

After warping, one fusion image, also called the proteome map, is created. It contains all the information about all the spots detected during the experiment. The software must find the spot positions, their surrounding boundaries and determine their quantities on the base of this proteome map. Due to the complex pattern of gel images and because of the existence of weak or merged spots and noises, this automated spot detection may not be satisfactory. Manual interventions are permitted, but different operators may have different ideas about the "correct" spot shapes, and such intervention may result in poor reproducibility between different operators. Correct background subtraction is also very important and has a critical influence on the spot number and subsequent quantitation.

Normalization of spot quantities is the next critical step, and its aim is to mitigate systemic differences between images that may be caused by sample loading and staining efficiency. To compare certain spot intensity between the gels, a step called spot matching has to be performed. Each spot on a given gel is mapped to the corresponding spot on another. An approach to achieve this goal is based on a spot consensus pattern. Here, the boundaries of spots from the proteome map are transferred back to the original image using transformations that were produced during warping. Those boundaries are then remodeled on the original gel images to fit to their gray-level distributions. The spot quantities are then calculated by summing up the intensities of all the pixels inside the spot boundary. After spot matching and quantitation, certain statistical tests must be applied to indicate the changes that seem to be relevant from a biological point of view.

Although at first glance software gel analysis seems well established and automated, the procedure considering many factors may be time-consuming, demands a skilled operator and may be an important source of variance due to its imperfection and various algorithms applied. To minimize variations, several gels should be run for the same sample (at least triplicates) to achieve improved statistical analysis.

7.2.7 Difference Gel Electrophoresis

DIGE is a type of 2D gel electrophoresis in polyacrylamide gel [11,12] where different samples are stained with different fluorescent dyes and then separated simultaneously on the same gel. This approach removes potential variance between gels. It is known that only a small amount of proteins will show changes in expression, or will be posttranslationally modified due to the examined process. The rest of remaining molecules will be unmodified and theoretically they should be localized at the same positions on the gels.

Staining different samples with different fluorescent dyes and their separation on the same gel enables both control and experimental samples to undergo identical conditions during the separation process. In that way it is possible to eliminate gel-to-gel variations between the control and experimental samples. It has to be stressed that DIGE does not eliminate gel comparison. There is still a need to compare the gels that represent statistical repetitions. Introduction of internal standard may help overcome this obstacle. Additionally, we may decrease the number of gels run during one experiment; for three control and three experimental samples we need six gels in the classical approach and only three in case of DIGE. This enables faster and more reproducible identification of the differences in protein expression between two samples (for the scheme of analysis, see Fig. 7.2.5.).

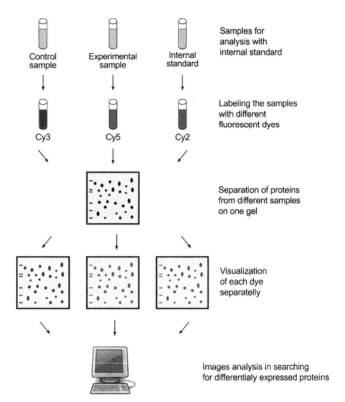

Figure 7.2.5 Scheme of DIGE analysis.

7.2.8 Fluorescent Dyes Used in Difference Gel Electrophoresis

Fluorescent dyes used for protein separation on the same gel have to meet several conditions:

- They must possess the same MW and charge to assure that the same proteins stained with different dyes will be found at the same position on the gel.
- They must replace the charge characteristic for the amino acid residue to which they are bound.
- They must possess a different range of absorption and emission, which makes possible the observation of different proteins labeled with different fluorescent dyes.

There are three cyanine dyes used in DIGE (Table 7.2.2): Cy2, Cy3 and Cy5. They possess a broad

Table 7.2.2 Different dyes used for dige analysis

Dye	Absorption (nm)	Emission (nm)	Sensitivity (ng) (minimal labeling)
Cy2	488	520	0.075
Cy3	532	580	0.025
Cy5	633	670	0.025

dynamic range (more than 3.6 orders of magnitude) and are characterized by linearity and sensitivity (for minimal labeling: Cy2: 0.075 ng; Cy3: 0.025 ng; Cy5: 0.025 ng; and for saturation labeling: below 15 pg). Proteins labeled with Cy dyes may be analyzed and identified by mass spectrometry.

7.2.9 Internal Standard

Having three fluorescent dyes permits separating three samples on the same gel. One of these samples may be replaced by an internal standard. In this approach, the internal standard is the mixture of equal amounts of both samples taken for comparison, stained with one fluorescent dye (usually Cy2). This means that every protein from both experiments will be present and visualized. This system provides accurate quantitation and eliminates variations between gels, which leads to significantly increased accuracy.

Depending on the amount of material available for the analysis, it is possible to apply two different modes of labeling: minimal and saturation. Minimal labeling may be used for samples containing 50 μg of protein (eg, tissue or cell culture). For this purpose, all three Cy dyes may be used, and they bind to the molecules through lysine residue. The analysis is facilitated by using an internal standard (labeled with Cy2). Saturation labeling is applied for very small amounts of sample, even as low as 5 μg. Only

Cy3 and Cy5 may be used in saturation labeling, and in this case they react with cysteine residues of the proteins. There is no possibility, however, to include an internal standard.

7.2.10 Pros and Cons of Difference Gel Electrophoresis

DIGE still has some limitations characteristic of 2D gel electrophoresis, such as the problem with separation of hydrophobic proteins. Additionally, it may only be used when the proteins contain accessible lysine (for minimal labeling) or cysteine residues (for saturation labeling). In general, DIGE is much more sensitive and reliable and offers a broader dynamic range in comparison with classical gel staining. Additionally, it shows better reproducibility of the results, due to the fact that several samples may be separated on the same gel. The possibility of using an internal standard makes matching between repetitive gels easier and allows for obtaining better accuracy during quantitative analysis. The fact that a lower number of gel repeats is necessary for statistical purposes is also an asset.

References

[1] Beranova-Giorgianni S. Proteome analysis by two-dimensional gel electrophoresis and mass spectrometry: strengths and limitations. TrAC-Trends Anal Chem 2003;22(5): 273−81.

[2] Fey SJ, Larsen PM. 2D or not 2D. Two-dimensional gel electrophoresis. Curr Opin Chem Biol February 2001;5(1):26−33.

[3] Garbis S, Lubec G, Fountoulakis M. Limitations of current proteomics technologies. J Chromatogr A June 3, 2005;1077(1):1−18.

[4] Gorg A, Weiss W, Dunn MJ. Current two-dimensional electrophoresis technology for proteomics. Proteomics December 2004;4(12):3665−85.

[5] Kraj A, Silberring J. Two-dimensional gel electrophoresis. Hoboken (NJ): John Wiley & Sons, Inc.; 2008.

[6] Patton WF. A thousand points of light: the application of fluorescence detection technologies to two-dimensional gel electrophoresis and proteomics. Electrophoresis April 2000;21(6):1123−44.

[7] Bierczynska-Krzysik A, Kang SU, Silberrring J, Lubec G. Mass spectrometrical identification of brain proteins including highly insoluble and transmembrane proteins. Neurochem Int August 2006;49(3):245—55.

[8] Young DA. Advantages of separations on "giant" two-dimensional gels for detection of physiologically relevant changes in the expression of protein gene-products. Clin Chem December 1984;30(12 Pt 1):2104—8.

[9] Berth M, Moser FM, Kolbe M, Bernhardt J. The state of the art in the analysis of two-dimensional gel electrophoresis images. Appl Microbiol Biotechnol October 2007;76(6):1223—43.

[10] Daszykowski M, Stanimirova I, Bodzon-Kulakowska A, Silberring J, Lubec G, Walczak B. Start-to-end processing of two-dimensional gel electrophoretic images. J Chromatogr A July 27, 2007;1158(1—2):306—17.

[11] Marouga R, David S, Hawkins E. The development of the DIGE system: 2D fluorescence difference gel analysis technology. Anal Bioanal Chem June 2005;382(3):669—78.

[12] Unlu M, Morgan ME, Minden JS. Difference gel electrophoresis: a single gel method for detecting changes in protein extracts. Electrophoresis October 1997;18(11):2071—7.

8

QUANTITATIVE MEASUREMENTS IN PROTEOMICS: MASS SPECTROMETRY

A. Drabik
AGH University of Science and Technology, Krakow, Poland

J. Silberring
AGH University of Science and Technology, Krakow, Poland;
Polish Academy of Sciences, Zabrze, Poland

CHAPTER OUTLINE

8.1 Introduction   146
8.2 Absolute Quantitation   146
8.3 Relative Quantitation in Proteomics   149
    8.3.1 Gel-Based Quantitative Proteomics   149
    8.3.2 Gel-Free Quantitative Proteomics: Isotope-Coded Affinity
        Tagging   149
    8.3.3 N-Terminal Labeling   150
    8.3.4 C-Terminal Labeling   152
    8.3.5 Labeling of Definite Amino Acid-Containing
        Peptides   153
    8.3.6 Metabolic Labeling   154
    8.3.7 Label-Free Techniques   155
8.4 Summary   157
Acknowledgments   158
References   158

Proteomic Profiling and Analytical Chemistry. http://dx.doi.org/10.1016/B978-0-444-63688-1.00008-2

8.1 Introduction

To meet the demands of monitoring and understanding biological processes, the "-omics" techniques, including proteomics, have evolved into elaborated tools of quantitative measurements. Therefore we expect that proteomics will advance our knowledge by direct and precise measurements of the levels of gene products present in a given state of a biological system. It is important to note that quantitative proteomics consists of two processes: protein identification and protein quantitation. While these two processes go hand-in-hand for many of the abundant proteins identified at the level of high confidence with multiple unique peptides, problems start to arise when our goal is quantitation of low-abundant proteins or those that are represented by few well ionizing and fragmenting peptides. How reliable is quantitation if in one sample there is only one peptide identified with medium confidence and in a counterpart sample there are two peptides identified, one with low and one with high confidence? Following output of iTRAQ analysis using ProteinPilot, we may set a threshold and reject all peptides that have been identified with 66% or less confidence. In many instances, this is the only peptide for one condition, and if it is filtered out it will give value "0" and thus the ratio cannot be calculated. Absolute quantitation based on spiking in known amounts of peptides isotopically labeled helps in quantitation changes in low-abundant proteins [1]. Nevertheless, quantitation becomes much more complicated when we try to quantitate peptides with posttranslational modifications.

8.2 Absolute Quantitation

The absolute signal intensity of the ion measured using a mass spectrometer (MS) does not always directly correlate with the abundance of peptide present in the analyzed sample. This is due to variability in ionization and the presence of other ions from high-complexity samples that may interfere with the signal in MS. Therefore, a reliable internal standard is required to normalize quantitative changes among different MS measurements. An ideal standard should behave identically in MS and be different in mass. Thus, the best internal standard for a peptide is an identical peptide labeled with stable isotope(s).

Several analytical approaches have been developed to meet the demand for absolute quantitation. One approach is the Protein Standard for Absolute Quantification (PSAQ™) [2], which utilizes full-length isotope-labeled proteins as isotope-dilution standards for MS-based quantification of target proteins. Strengths of this method include highly accurate quantification in extensively prefractionated samples [3], elimination of differences in digestion yields between the internal standard and the target protein [3,4], and providing the largest sequence coverage for quantification, thus providing increased detection specificity and measurement robustness [3]. This approach has been further adjusted by combining with immunocapture of proteins of interest for absolute quantitation [5].

Another strategy to absolute quantification of proteins, called QconCAT, utilizes a synthetic gene designed to encode all proteotypic peptides of the sample protein mixture. This gene is expressed in the medium with isotope-labeled amino acids, and a QconCAT synthetic polypeptide labeled with stable isotopes is purified and serves as an internal control in mass spectrometry analysis [6].

An absolute quantitation (AQUA) is based on spiking in AQUA synthetic peptides using fully labeled 98 atom% ^{13}C and 98 atom% 15N-enriched amino acids (one labeled amino acid per peptide) principles of single-reaction monitoring (SRM) in tandem mass spectrometry for quantitation of peptides [7].

Each of these methods has strengths and limitations, which need to be carefully considered before applied the method to any proteomic experiment (Table 8.1).

Table 8.1 Summary of different quantitative proteomic techniques

Labeling technique	Advantages	Drawbacks
Gel-based quantitative proteomics DIGE/D_3 acrylamide	Quantitation based on protein/peptide levels.	Modest dynamic range (up to 4 orders of magnitude). Limitation in pI and hydrophobic properties of separated proteins.
ICAT	Subproteome analysis (only cysteine containing peptides) allows for less abundant protein identification.	Observed shifts in chromatographic peaks.
N-terminal labeling NIT/acetylation iTRAQ	Enable quantitation of complex biological samples.	Tags in low *m/z* range prevent quantitation using ion-trap instruments (cutoff limit).
C-terminal derivatization Methylation/D_2O tryptic digestion	Simple nature of the reaction without affecting any biological properties.	Esterification is not specific to the carboxyl terminus only. Small mass difference between analyzed peptides complicates quantitation with low-resolution mass spectrometers.
Targeted amino acid labeling NBSCI/MCAT	Subproteomic quantitation enables low-abundant protein identification.	Internal standard signal is chemically distinct from the labeled sample.
Metabolic labeling ^{15}N/SILAC	Analysis of metabolic pathways is possible.	Provides analysis of samples from cell cultures only (at least 5 passages to incorporate isotopic labels).
Label-free	Lower costs, sample preparation protocol simple (reducing the incorporation of isotopic label reaction with its limited efficiency).	Variations in sequential analysis signal.

8.3 Relative Quantitation in Proteomics

8.3.1 Gel-Based Quantitative Proteomics

The traditionally carried out and still commonly used approaches for quantitative proteomics are the gel-based methods, such as differential gel electrophoresis (DIGE), followed by identification using MS. In this procedure, proteins are separated by 2D gel electrophoresis (2DE) and quantified based on the intensity of the protein spots, followed by MS identification. Moreover, deuterated acrylamide labeling of proteins during alkylation of cysteine groups was proposed by Sechi et al. [8]. After mixing both acrylamide and D_3-acrylamide-labeled samples at a 1:1 ratio, the mixture is loaded on the electrophoretic gel. The bands are excised, digested and analyzed. Isotopic distribution of the cysteine-containing peptides results in formation of two isotopic signals $3\,m/z$ apart. Quantification is accomplished by comparing the intensities of the "light" and deuterated components. This provides relative but not absolute quantitation. An additional benefit of this approach is stabilization of cysteines. Proteomic analysis by 2DE/MS is restricted due to the limitations of the method for certain species, such as membrane proteins, excessively large or small proteins, and very acidic or basic proteins. Moreover, one has to keep in mind that some proteins co-migrate in 2D gels, which can produce unreliable quantitative results.

8.3.2 Gel-Free Quantitative Proteomics: Isotope-Coded Affinity Tagging

Development of isotope-coded affinity tagging (ICAT) was a significant step in relative quantitation [9]. In vitro stable isotope reagent consists of a biotin affinity tag for selective purification, a linker that incorporates stable isotopes ^{1}H and ^{2}H, and iodoacetamide (IA) reactive group specifically reacting with cysteinyl thiols (Fig. 8.1). Proteins obtained from two different samples are separately labeled at their cysteine residues with either light or heavy ICAT

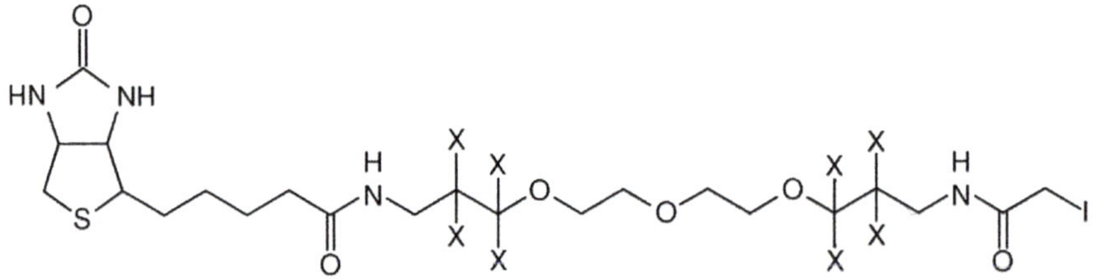

Figure 8.1 Structure of isotope-coded affinity tag.

reagent, respectively. The "light" and "heavy" samples are combined, digested by proteolytic enzyme, and the resulting peptides fractionated by multidimensional chromatography and quantitatively analyzed by MS. Ion intensity ratios between the heavy and light forms of a specific peptide, with a mass shift of 8 Da, are indicative of their relative abundance. The ICAT technique has since been improved and now it contains an acid-cleavable linker that allows for removal of the large affinity tag prior to MS analysis [10]. Another refinement of ICAT technology introduced $^{12}C/^{13}C$ isotopes to prevent chromatographic peaks shifts (hydrogen and deuterium labeling of peptides affects retention times). Finally, a solid-phase variant of the ICAT procedure has been performed for simpler enrichment of target peptides, which leads to the automatic, selective purification of cysteine-containing peptides, thus significantly reducing sample complexity for simple detection of the low-abundance compounds in biological samples.

Those findings captured the attention of many researchers and initiated significant progress in development of similar methods. Important contributions made over the past few years for quantitative proteomics rely on peptide modifications at the N-terminus.

8.3.3 N-Terminal Labeling

N-terminus peptide labeling based on modification using a variety of reagents that are commercially available (eg, O-methylisourea, acetic anhydride, propionic anhydride) is termed N-terminal isotope-encoded tagging (NIT) and is a relatively straightforward procedure [11]. However, it is problematic to reliably control the selectivity of the reaction, because the ε-amino groups of lysine residues can also be stably modified by reagents that target the N-terminus. Guanidination of lysine with O-methylisourea after enzymatic digestion with trypsin and prior to acetylation of the N-terminus might be one solution. N-terminal acetylation by derivatization with acetic anhydride and deuterated acetic anhydride [12] was performed to quantify peptides using electrospray and MALDI ionization mass spectrometry [13]. Modification of peptides is completed after enzymatic cleavage with a 1:1 mixture of light and heavy acetyl group. All N-terminal fragment ions retain the isotopic label that differs by 3 m/z and the N-terminal sequence can be revealed directly from the mass spectrum. It was shown that this derivatization procedure performed under restricted conditions does not affect the ε-amino groups of lysine residues.

Another approach is based on amine group labeling and referred to as isobaric tags for relative and absolute quantitation (iTRAQ) [14]. The iTRAQ reagent consists of a reporter group, a balance group, and a peptide reactive group (Fig. 8.2). The reporter group is a tag with masses of 113, 114, 115, 116, 117,

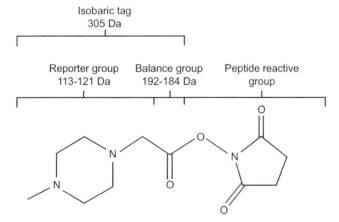

Figure 8.2 Structure of isobaric tags for relative and absolute quantitation.

118, 119, and 121 Da, depending on various isotopic combinations of $^{12}C/^{13}C$, $^{14}N/^{15}N$, and $^{16}O/^{18}O$ in each individual reagent. The balance group differs in mass to ensure the combined mass of the reporter and balance groups remain constant and equal to 305 Da for all eight reagents, and 145 for the four-plex. Therefore peptides labeled with different isotopes are isobaric and are indistinguishable during chromatographic separation. During collision-induced dissociation (CID), the reporter ions are truncated from the backbone peptides displaying distinct masses between 113 and 121 Da, respectively. The intensities of theses fragments are subsequently used for quantitation of the individual peptides representing various protein sets. In contrast to other stable isotope labeling techniques, iTRAQ quantitates the relative peptide abundance from fragmentation spectra. Absolute quantitation can be achieved after spiking previously labeled standard protein into the analyzed sample. This approach allows for simultaneous labeling of up to eight samples during single experiment, which is what makes iTRAQ eight-plex very valuable for complex studies. Originally, iTRAQ was devoted to measurements by the MALDI TOF/TOF method, but more recently is being measured by quadrupole-based analyzers. The major obstacle of this method is the inability to use quadrupole ion-trap instruments as the ion trap has a cut-off at the low m/z range.

8.3.4 C-Terminal Labeling

Analogous to the labeling of the N-terminal amine groups, the carboxyl termini can also be derivatized with the use of stable isotopes. The first attempts of C-terminal derivatization introduced stable isotopes to the C-terminus by esterification using $^{1}H/^{2}H$ methanol [15]. A main disadvantage of this approach was that esterification is not specific to the carboxyl terminus only, as aspartate and glutamate residues can also be labeled. Nevertheless, this was a first step to a technique where chemical introduction of isotopes, such as ^{18}O, to the C-termini was linked with the enzymatic reaction, using proteolytic

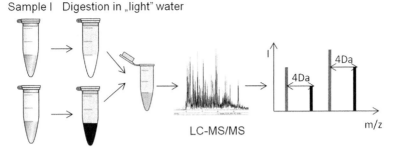

Sample I Digestion in „light" water

LC-MS/MS

Sample II Digestion in „heavy" water

Protein identification and quantitation

Figure 8.3 Quantitative approach based on proteolytic [18]O labeling.

enzymes [11]. Labeling of the peptides with the isotopic form of oxygen occurs along with the action of proteolytic enzymes, such as trypsin, chymotrypsin, and Glu-C. During trypsin digestion, in the presence of ^{18}O water, the heavy isotope is incorporated into the carboxyl terminus (Fig. 8.3). The relative quantity of peptides is determined by the ratio of ion intensities between the $^{16}O/^{18}O$ labeled species, which is measured by MS. The resulting mass difference of 2–4 Da does not alter chromatographic separation, and thus heavy and light peptides elute at the same retention time. During MS analysis, only the Y ions will retain the isotope label, thus making identification and interpretation more comprehensive. The main benefits of this process are simplicity and the ability to target the commonly present groups without affecting any biological properties. Nonetheless, the small mass difference between heavy and light peptide versions complicates quantitation, especially with acquisition on low-resolution mass spectrometers. Furthermore, spontaneous exchange of ^{18}O to ^{16}O can occur due to continuous enzyme activity which can also complicate peptide quantitation and determination of isotope ratio. To avoid this effect, the enzyme should be inactivated by lowering pH or temperature.

8.3.5 Labeling of Definite Amino Acid-Containing Peptides

Another effort toward development of new labeling methods was made to target specific amino acid-containing peptides (similarly to the ICAT technique). Tryptophan-containing peptides are modified using $^{12}C/^{13}C$ and termed isotopically differentiated 2-nitrobenzenesulfenyl chloride (NBSCI) [14]. The labeled tryptophan residues are enriched by Sephadex™ chromatography, which is based on the increased hydrophobic properties of the attached NBSCI particle. The enriched peptides are then identified by MS and quantified by the ratio of the ion intensities arising from the isotope peaks. Furthermore, lysine-containing peptides can also be quantified by a method termed mass-coded abundance tagging (MCAT) [16]. The MCAT method specifically labels the lysine ε-amino groups by guanidinylation obtained by O-methylisourea reaction. This approach is not a strict isotope labeling, as the internal standard is chemically distinct from the labeled sample by more than just isotopic atoms, and the molecule, as well as physical properties between samples, may cause the loss of accuracy of quantitation during the separation process.

8.3.6 Metabolic Labeling

In vivo labeling approaches involve metabolically incorporated stable isotopes into proteins of the intact cells cultured in special media. In the metabolic labeling approach, isotopically defined media are used to culture two or more biologically different samples. Originally, ^{15}N-containing media were used to incorporate an isotopic form of nitrogen into the proteins in intact plants, microorganisms or cell cultures [17]. The mass shift that occurred after this treatment allowed for protein ratio determination based on the ion intensities of the isotopically labeled peptide pairs. However, the mass shift of unknown peptides cannot be predicted. The most critical drawbacks of this method are incomplete nitrogen incorporation and difficulties in data processing. This was the main reason for development of another approach, stable isotope labeling with amino acids in cell culture (SILAC) [18]. In this method, proteins are labeled by growing cells in the media containing isotopically modified amino acids including $^{2}H_4$-leucine, $^{2}H_4$-lysine, $^{13}C_6$-lysine, $^{13}C_6$-$^{15}N_2$-lysine,

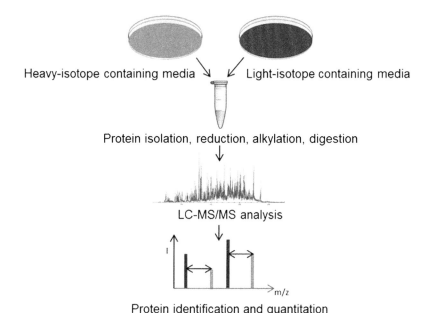

Heavy-isotope containing media   Light-isotope containing media

Protein isolation, reduction, alkylation, digestion

LC-MS/MS analysis

Protein identification and quantitation

Figure 8.4 Experimental approach for SILAC quantitation.

$^{13}C_6$-$^{15}N_4$-tyrosine, $^{13}C_6$-arginine, and $^{13}C_6$-$^{15}N_4$-arginine. The principle of SILAC is that it utilizes modified amino acids that are essential for the cell culture (Fig. 8.4) and can be metabolically incorporated into the cellular proteome. The particular advantage of using a labeled Lys/Arg mixture emerges from trypsin digestion, which cleaves at the C-terminal to Lys/Arg, thus typically labeling a single isotope on each peptide, allowing for effective identification and quantitation by MS. This technique gained popularity because of the easily calculated mass shift, which can be analyzed by MS. Nevertheless, the drawback of this method is in the difficulty to analyze biological samples, for example tissues or body fluids. Additionally, this procedure requires long exposure and incubation time in cell culture (at least five passages) to fully incorporate isotopic labels.

8.3.7 Label-Free Techniques

Label-free quantitation is an attractive approach for three major reasons. First, the variability that chemical labeling/tagging may introduce is eliminated. Second, chemical and metabolic tags are usually expensive. Third, the time for sample preparation is significantly reduced by elimination of numerous steps. While the relative abundance of chemical or metabolic tags can be easily measured, label-free quantitation must rely on other parameters such as peptide or spectral count, which also has inherent limitations. Other issues associated with label-free MS-based quantitation are sequence coverage and the degree of complex sample fractionations prior to analysis in a mass spectrometer. All of these issues and limitations need to be carefully considered before a decision of the optimal approach for a specific experiment is made.

Label-free quantitative protein profiling based on LC-MS stems out of the premise that the relative intensities of extracted ion chromatograms is a measure of relative abundance of peptides, which is related to concentration of protein in a complex sample. Based on the observed linear correlation between the peak area of measured peptides and their abundances, they can be quantified through the signal intensity ratio of their corresponding counterparts compared among MS runs. Alignment of data is three-dimensional, which includes elution time (x axis), m/z value (y axis) and signal intensity (z axis). Systematic error introduced by variables in any of these three data points representing a peptide will be detrimental to the overall data set, generating false-positive results, and subsequently the entire experiment may fail. A major disadvantage of peptide quantitation by the signal intensity is that it often includes experimental variation (retention times for all counterparts must be very close) and signal-to-noise, which can affect the accuracy. Excellent chromatographic conditions are necessary to minimize shifts in elution time of the same peptide between analytical runs (replicates). High-quality resolution and mass accuracy are another two parameters that contribute to the success of label-free quantitative proteomic experiments. Increasing the number of replicates—usually three are the minimum—will not help in accurate comparisons if the sample preparation and condition of instrumentation are not superb. Elevated background noise from samples and reagents may further impair the outcome at the protein identification and quantitation levels. Another issue for consideration is the coelution of peptides, which is an inherent characteristic of complex samples. Alternatively, spectral counting can also be utilized as a label-free quantitation method, where the number of MS/MS spectra assigned to each protein is used [19].

Analytical measurements of samples designed for label-free MS-based quantitative approaches are only the first step, and the remaining actions that follow depend on algorithms. There are statistical tools available for analyzing data obtained from multiple LC-MS analyses that allow reducing variations between runs. Development of such software packages has accelerated in recent years and there are several commercially available, such as Progenesis LC-MS (NonLinear Dynamics, Durham, NC, USA), ProteinLynx (Waters, Milford, MA, USA), Elucidator (Rosetta Inpharmatics, Seattle, WA, USA), Decyder MS (GE Healthcare, Piscataway, NJ, USA), and SIEVE (Thermo Fischer Scientific, San Jose, CA, USA). Several algorithms were formally compared against label based quantitation techniques such as iTRAQ [20,21]. Algorithms are also parts of software packages that can be multimodal and used for processing of label-free or labeled MS data such as MaxQuant [22]. For a more in-depth overview, we direct our readers to several excellent publications [23,24].

8.4 Summary

By means of constant progress in MS technologies, quantitative proteomics has grown amazingly in the past few years. The in vivo stable isotope labeling technology provides a reliable and accurate approach to measure protein abundances. The limitation of this technique is that it can only be applied to cultured cells, and thus is not suitable for applications to tissues or body fluids, which are of particular interest for medical research. The in vitro labeling methods, including commercially available ICAT and iTRAQ methods, can be applied to a wide variety of biological samples. The ICAT technique focuses on cysteine-containing peptides and can be used for global proteome quantitation, while iTRAQ, which is designed for amino-termini, is especially useful for protein quantitation in less complex samples. There is still a need for new proteomics strategies for quantitative measurements that provide crucial knowledge on dynamic changes during multiple cellular processes that can be employed and integrated with various separation techniques, such as 1-DE, 2-DE, and LC. Special concern should be devoted to the selection of appropriate quantitation approaches according to the needs of the sample quality and experimental design.

Acknowledgments

The authors would like to thank the NSC, as this chapter was partially supported by grants from the Polish National Science Center (grant numbers 3744/B/H03/2011/40 and 3048/B/H03/2009/37).

References

[1] Bronstrup M. Absolute quantification strategies in proteomics based on mass spectrometry. Expert Rev Proteomics December 2004;1(4):503−12.

[2] Jaquinod M, Trauchessec M, Huillet C, Louwagie M, Lebert D, Picard G, et al. Mass spectrometry-based absolute protein quantification: PSAQ strategy makes use of "noncanonical" proteotypic peptides. Proteomics April 2012;12(8):1217−21.

[3] Brun V, Dupuis A, Adrait A, Marcellin M, Thomas D, Court M, et al. Isotope-labeled protein standards: toward absolute quantitative proteomics. Mol Cell Proteomics December 2007;6(12):2139−49.

[4] Janecki DJ, Bemis KG, Tegeler TJ, Sanghani PC, Zhai L, Hurley TD, et al. A multiple reaction monitoring method for absolute quantification of the human liver alcohol dehydrogenase ADH1C1 isoenzyme. Anal Biochem October 1, 2007;369(1):18−26.

[5] Dupuis A, Hennekinne JA, Garin J, Brun V. Protein Standard Absolute Quantification (PSAQ) for improved investigation of staphylococcal food poisoning outbreaks. Proteomics November 2008;8(22):4633−6.

[6] Beynon RJ, Doherty MK, Pratt JM, Gaskell SJ. Multiplexed absolute quantification in proteomics using artificial QCAT

proteins of concatenated signature peptides. Nat Methods August 2005;2(8):587−9.

[7] Tan CT, Croft NP, Dudek NL, Williamson NA, Purcell AW. Direct quantitation of MHC-bound peptide epitopes by selected reaction monitoring. Proteomics June 2011;11(11):2336−40.

[8] Ranish JA, Brand M, Aebersold R. Using stable isotope tagging and mass spectrometry to characterize protein complexes and to detect changes in their composition. In: Sechi S, editor. Quantitative proteomics by mass spectrometry. Totowa (NJ, USA): Humana Press; 2007.

[9] Gygi SP, Rist B, Gerber SA, Turecek F, Gelb MH, Aebersold R. Quantitative analysis of complex protein mixtures using isotope-coded affinity tags. Nat Biotechnol October 1999;17(10):994−9.

[10] Elliott MH, Smith DS, Parker CE, Borchers C. Current trends in quantitative proteomics. J Mass Spectrom 2009;44:1637−60.

[11] Ong SE, Mann M. Mass spectrometry-based proteomics turns quantitative. Nat Chem Biol October 2005;1(5):252−62.

[12] Morris HR, Panico M, Barber M, Bordoli RS, Sedgwick RD, Tyler A. Fast atom bombardment: a new mass spectrometric method for peptide sequence analysis. Biochem Biophysical Res Commun July 30, 1981;101(2):623−31.

[13] Noga MJ, Asperger A, Silberring J. N-terminal H_3/D_3-acetylation for improved high-throughput peptide sequencing by matrix-assisted laser desorption/ionization mass spectrometry with a time-of-flight/time-of-flight analyzer. Rapid Commun Mass Spectrom-RCM 2006;20(12):1823−7.

[14] Iliuk A, Galan J, Tao WA. Playing tag with quantitative proteomics. Anal Bioanal Chem January 2009;393(2):503−13.

[15] Caroll KM, Lanucara F, Eyers CE. Quantification of proteins and their modifications using QconCAT technology. In: Westerhoff H, Verma M, Jameson D, editors. Methods in systems biology. USA: Academic Press; 2011. p. 113−30.

[16] Oeljeklaus S, Barbour J, Meyer HE, Warscheid B. Mass spectrometry-driven approaches to quantitative proteomics and beyond. In: Whitelegge JP, editor. Protein mass spectrometry. Amsterdam (The Netherlands): Elsevier; 2008. p. 411−49.

[17] Lanucara F, Eyers CE. Mass spectrometis-based quantitative proteomics using SILAC. In: Westerhoff H, Verma M, Jameson D, editors. Methods in systems biology. USA: Academic Press; 2011. p. 133−48.

[18] Huttlin EL, Hageman AD, Susseman MR. Metabolic labeling approaches for the relative quantification of proteins. In: Whitelegge JP, editor. Protein mass spectrometry. Amsterdam (The Netherlands): Elsevier; 2008. p. 479−515.

[19] Ross PL, Chen X, Toro E, Britos L, Shapiro L, Pappin D. Multiplexed quantitative proteomics using mass spectrometry. In: Whitelegge JP, editor. Protein mass

spectrometry. Amsterdam (The Netherlands): Elsevier; 2008. p. 449–67.

[20] Patel VJ, Thalassinos K, Slade SE, Connolly JB, Crombie A, Murrell JC, et al. A comparison of labeling and label-free mass spectrometry-based proteomics approaches. J Proteome Res July 2009;8(7):3752–9.

[21] Trudgian DC, Ridlova G, Fischer R, Mackeen MM, Ternette N, Acuto O, et al. Comparative evaluation of label-free SINQ normalized spectral index quantitation in the central proteomics facilities pipeline. Proteomics July 2011;11(14):2790–7.

[22] Cox J, Mann M. MaxQuant enables high peptide identification rates, individualized p.p.b.-range mass accuracies and proteome-wide protein quantification. Nat Biotechnol December 2008;26(12):1367–72.

[23] Mueller LN, Brusniak MY, Mani DR, Aebersold R. An assessment of software solutions for the analysis of mass spectrometry based quantitative proteomics data. J Proteome Res January 2008;7(1):51–61.

[24] Matthiesen R, Carvalho AS. Methods and algorithms for relative quantitative proteomics by mass spectrometry. Methods Mol Biol 2010;593:187–204.

9

SWATH-MS: DATA ACQUISITION AND ANALYSIS

K. Frederick and P. Ciborowski

University of Nebraska Medical Center, Omaha, NE, United States

CHAPTER OUTLINE

9.1 Introduction   161
9.2 Tandem Mass Spectrometry for Quantitative Proteomics   162
  9.2.1 Data-Dependent Acquisition   163
  9.2.2 Data-Independent Acquisition   165
9.3 SWATH-MS Data Acquisition   165
  9.3.1 Spectral Library   166
  9.3.2 Targeted Data Extraction   168
9.4 Overview of SWATH-MS Data Analysis   168
  9.4.1 MarkerView: Data Normalization and Principal Component Analysis   169
  9.4.2 Statistical Analysis Using Z-Transformation   169
  9.4.3 Spectronaut   170
  9.4.4 Skyline   170
  9.4.5 OpenSWATH   171
9.5 Summary   171
References   172

9.1 Introduction

Along with the constant technological development of mass spectrometers, the quest for obtaining more information from the analysis of any proteomic sample never ends. Moreover, the proteome is dynamically changing with respect to time.

Proteomic Profiling and Analytical Chemistry. http://dx.doi.org/10.1016/B978-0-444-63688-1.00009-4

Monitoring such changes requires analyzing numerous samples, which is further multiplied by the number of replicates (biological and technical). Multiple replicates might be a time-prohibitive proposition if samples are not readily available and the preparation is lengthy, costly and laborious. As such, the ideal analysis of a proteomic sample will be the result of an experimental run that extracts the highest amount of information possible.

Proteomic profiling based on data-independent acquisition (DIA) mass spectrometric quantification, whether global profiling or more focused, is a substantial step toward addressing these issues. DIA analysis does not require any kind of labeling (chemical or metabolic) and, unlike in data-dependent acquisition (DDA), all the information from a sample is recorded. DIA-derived data depends on the mass range and window size selected for MS/MS analysis, while in DDA mode only a limited number of precursor ions are selected according to the intensities of precursor ions. One example of DIA analysis is SWATH-MS, a proteomic platform developed by Sciex, Inc. (www.sciex.com). However, the application of this platform comes with the price of how data are analyzed and how much important information we are able to extract from any dataset acquired experimentally. Although this chapter focuses on SWATH-MS, there are a variety of DIA-based platforms available currently.

SWATH-MS data acquisition integrates data-dependent and data-independent approaches for simultaneous protein identification and quantification. It is important to note that data extraction from SWATH-MS DIA analysis requires a reference library of spectra generated in DDA mode. Fig. 9.1 outlines the discussion of the SWATH-MS workflow as presented in this chapter.

9.2 Tandem Mass Spectrometry for Quantitative Proteomics

One of the major technological pillars of proteomics research is mass spectrometry [1]. Mass spectrometry (MS) is a technique in which proteins and

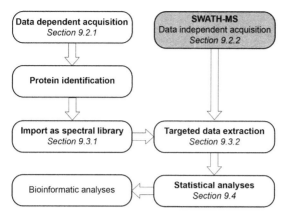

Figure 9.1 Overview of SWATH-MS workflow. Data-dependent acquisition is used to create a spectral library. The library is used for targeted data extraction of SWATH-MS data-independent acquisition-derived data. Italics indicate sections of this chapter dedicated to more detailed explanation for each step in SWATH-MS analysis.

peptides as well as other types of molecules are ionized and analyzed based on their characteristic mass to charge (m/z) values [2]. Computer algorithms and databases are then used to match the ions to proteins or small molecules (Fig. 9.2A). Many quantitative MS platforms fall into the category known as tandem MS, which involves analyzing the ionized sample by sequential fragmentation of ions in a gas phase. In the first step, MS1, the mass spectrometer performs a survey scan to generate a list of precursor ions, which are further sequentially fragmented and analyzed in the second step, MS2. Because many molecules, peptides in particular, may have different structures represented by identical or close to identical m/z values, tandem MS allows for precise identification of molecules being investigated.

9.2.1 Data-Dependent Acquisition

Data-dependent acquisition (DDA) is the classic form of tandem MS (Fig. 9.2B). In the first step of DDA, MS1, the mass spectrometer performs a survey scan to generate a list of precursor ions. That list is then used to identify the most abundant ions, which

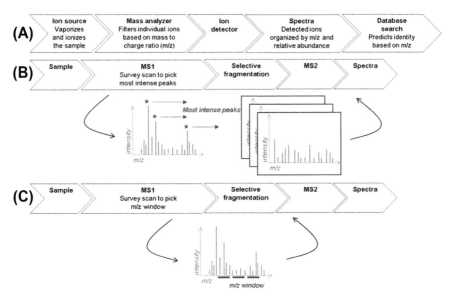

Figure 9.2 Tandem mass spectrometry for quantitative proteomics. (A) Basic schematic of mass spectrometry. (B) Data-dependent acquisition. Precursor ions are selected based on a predefined threshold. (C) Data-independent acquisition. Precursor ions are selected in an *m/z* window.

will be subjected to fragmentation in MS2, starting with the most abundant precursor ions. Precursor ions are typically selected above a predefined threshold value; however, other possible selection criteria may also be applied. The number of precursor ions to be fragmented is set by the investigator and may range from a few ions to 50 or more. Thus the final spectra depend on the characteristics of the data, as seen in the MS1 survey scan [3].

One major limitation of DDA is the speed of data acquisition. Low-abundant peptides might be eluted from in-line nano-LC column for several seconds and may not be available for fragmentation at the desired concentration. Samples with a dynamic range of protein concentrations pose an additional challenge. It has been shown that the most abundant protein in a sample is often at least two to three orders of magnitude higher in concentration than the majority of proteins present [4]. This can lead to fragmentation of background interference rather than ions from proteins present in the sample.

9.2.2 Data-Independent Acquisition

Recent advances in mass spectrometry of proteins and peptides have led to the development of the data-independent acquisition (DIA) approach (Fig. 9.2C). Instead of selecting specific precursor ions as a result of MS1 survey scan, all ions within a specific m/z region are fragmented in the second step (MS2). Sequential segments of m/z regions are fragmented, leading to increased proteomic coverage of the sample. These m/z windows are typically equivalent to 25 Da over 400 or 1500 m/z [3].

One of greatest challenges of quantitative proteomic profiling is the dynamic range of protein concentrations in a complex mixture. While contaminated proteins or impurities, as shown by Xie et al. [5], can be easily detected and quantified over several orders of magnitude, highly complex samples are posing additional challenges. Nevertheless, the results presented by these authors are promising for using DIA for quantification of posttranslationally modified proteoforms which might constitute a small fraction of the total pool of any given protein.

The improvements of MS instrumentation that have allowed for DIA are continuing to be matched on the analytic side. DIA-derived data is especially difficult due to the fact all the information in the sample is recorded, leading to a massive amount of information. The problem lies in how to process the raw data in such a way to increase the percentage of meaningfully extracted data, with 100 percent meaningful extraction as the ideal situation. Several groups are developing methodologies to analyze the output data from DIA experiments, aiming to improve m/z accuracy and the signal-to-noise ratio by averaging or clustering spectra in adjacent scans [6]. Thus, DIA is an evolving strategy that does not require detection or knowledge of precursor ions.

9.3 SWATH-MS Data Acquisition

One method of label-free quantification that has been commercialized by Sciex, Inc., is Sequential Windows Acquisition of All Theoretical Spectra-Mass Spectrometry (SWATH-MS) data acquisition, which

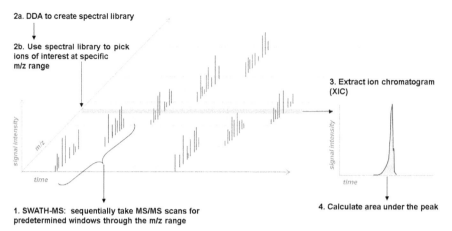

Figure 9.3 Targeted data extraction for SWATH-MS-based protein quantification. 1. Mass spectrometer steps through *m/z* windows in MS1, allowing all ions in that mass range to be analyzed in MS2. 2a. Mass spectrometer is run in data-dependent acquisition mode; data are searched in databases for protein identification as the reference spectral library. 2b. SWATH-MS data and DDA-MS library data are compared to focus analysis. 3. An extracted ion chromatogram (XIC) shows signal intensity versus time for a given *m/z*. 4. The area under the peak is calculated; this value is the ion quantification value.

is based on the concept of DIA [7−9]. SWATH-MS was developed with three major aims: to remove variability of chemical modifications, to increase the dynamic range of quantification, and to maximize the acquisition of all data that a sample may contain.

A concept of DIA acquisition using SWATH-MS technology is presented in Fig. 9.3. SWATH-MS data acquisition is distinctive as it incorporates the benefits of data-independent and data-dependent acquisition. Data are acquired using a mass spectrometer in DIA mode, but analysis requires a spectral library acquired using a mass spectrometer in DDA mode. The library of spectra is necessary for protein identification while the spectra acquired using SWATH-MS mode are used for quantification.

9.3.1 Spectral Library

SWATH-MS DIA quantitative analysis requires a library of spectra established in advance. The library is generated with the mass spectrometer in DDA mode and is used for targeted data extraction.

Typically, protein identification is performed on DDA-derived data using ProteinPilot software (current version 5.0) developed by Sciex, Inc. ProteinPilot analyzes the spectra using a variety of databases, including Swiss-Prot Uniprot Knowledgebase. FDR analysis can also be performed. This information is imported into PeakView software (current version 2.2), also developed by Sciex, Inc., to generate the library.

Comparison of DIA data to the spectral library is one of the first steps in SWATH-MS data analysis. If peptides or proteins are not in the library, they cannot be identified or quantified in the sample subjected to SWATH-MS data acquisition. It has been proposed that one universal library of spectra can be built and used by many, if not all, investigators, and such a library has been made available by Peptide Atlas (www.peptideatlas.org). While this may be the future approach, at this time it seems that a locally generated library is favored. One disadvantage of a universal library is the alignment of elution times of peptides from experimental DIA acquisition. A locally generated library can be built once and used for many SWATH-MS-based experiments, given the same LC gradient used to elute peptides inline with a mass spectrometer. This allows for retention times of eluted peptides to be within a similar range.

Another factor that might be related to control of the overall quality of proteomic experiments is the frequency of identified peptides in set of replicates. More specifically, we found that when using 15 sections of mouse brain tissue to generate a spectral library, approximately 8−12% of peptides were identified only once in those 15 independent DDA analytical runs (unpublished data). If these peptides are of low abundance due to poor ionization but belong to proteins represented by many other peptides, this concern might be of a lesser importance. However, if such peptides belong to protein represented by only a single additional peptide with a high-quality spectrum, this might impact quantification. A library of spectra can be expanded at any time by adding more information from DDA acquisitions, further increasing the draw of a locally generated library.

9.3.2 Targeted Data Extraction

SWATH-MS data acquisition creates a complex 3D map of m/z-intensity-retention time that cannot be searched using traditional database mechanisms [8]. Without the ability to use databases for automated searches, protein identification becomes an arduous task. The solution involves directing analysis toward the proteins found in the DDA-derived library of spectra. The spectral library is used to pick ions of interest. The signal intensities across time for a specific m/z window is compiled to create the extracted ion chromatogram (XIC) for each ion in the spectral library. The area under the XIC peak for ions is exported as the ion quantification value. Protein quantification values are calculated from the ion quantification values. The process of MS data acquisition is reversed by finding the sum of ion and peptide quantification values to calculate peptide and protein quantification values, respectively.

Targeted data extraction and spectral alignment of SWATH-MS data with the DDA-derived spectral library can be accomplished using PeakView software developed by Sciex, Inc. PeakView Software (current version 2.2) is a stand-alone software platform which, among other applications, has an intuitive interface for quantitative proteomics based on SWATH-MS data acquisition. PeakView enables exploration and interpretation of mass spectral data with tools for processing accurate mass data, structural interpretation, and batch analysis. It is compatible with Sciex, Inc. mass spectrometers supporting nominal mass (QTRAP) as well as accurate mass (TripleTOF data).

9.4 Overview of SWATH-MS Data Analysis

Regardless of the mode of proteomic data acquisition, whether it is DDA or DIA, rigorous statistical analysis must be performed to extract these results which show us statistically significant differences. Data processing, including statistical analyses, is a critical step before further analysis using bioinformatics. Output data provided by PeakView software (see the preceding section) can be further analyzed using one of many software packages for statistical analysis, some of which are open source and others of which are proprietary.

9.4.1 MarkerView: Data Normalization and Principal Component Analysis

MarkerView Software (current version 1.2.1) is a data analysis program developed by Sciex, Inc. This tool is designed for protein, peptide, and metabolomic biomarker profiling that allows for reviewing data acquired on all Sciex mass spectrometers to determine upregulation and downregulation of endogenous compounds in complex samples. MarkerView can also be used to analyze SWATH-MS data, including an exploration of the data using Principal Component Analysis.

Principal component analysis (PCA) is a mathematical technique used to reveal patterns and relationships between data points in a dataset. MarkerView performs PCA on a proteomic dataset by comparing multiple proteomic samples to reveal groupings of proteins or samples, which can be visualized to emphasize variation using a plot known as a Scores plot. The groups can be further explored using a plot known as the Loadings plot; proteins are grouped and colored according to upregulation or downregulation. MarkerView can additionally analyze the PCA groupings using a statistical t-test. Other tools and apps for proteomic data processing can be found as part of a collaborative effort. For example, Protein Expression Assembler and Protein Expression Workflow applications developed by Sciex, Inc., are available on Illumina's BaseSpace cloud computing site (https://basespace.illumina.com/home/prep).

9.4.2 Statistical Analysis Using Z-Transformation

Any quantitative proteomic data need to undergo thorough statistical analysis for identification of statistically significant differences, and this is independent from the type of technology used, ie, chemical, metabolic labeling or label-free. Although the sources might be different, each of these approaches have inherent variability. Because the raw intensity data (peak intensity, area under the peak, etc.) generated by mass spectrometry are inherently skewed, mass spectrometry data requires normalization to achieve the normal distribution required for parametric testing.

In our previous work [10] we proposed a z-score transformation of SWATH-MS data. First the proteomic data for each SWATH-MS experiment natural log (ln) is transformed and then normalized by the z-score, as described in Cheadle et al. [11]. This transformation minimizes distortions introduced from sample preparation and data acquisition. A paired sample z-test, which is conceptually equivalent to the paired sample t test, can be used to identify differences in protein expression between conditions while accounting for variation between biological replicates on a protein-by-protein basis.

9.4.3 Spectronaut

BiognoSYS (http://www.biognosys.ch/next-generation-proteomics.html), founded in 2008, is a company providing services and products for protein discovery and quantification. Their flagship software is Spectronaut, which was specifically developed for the analysis of quantitative proteomics based on DIA. Spectronaut software can analyze data acquired from SWATH or HRM (Hyper Reaction Monitoring). The latter is the Biognosys next-generation proteomics technology based on DIA. Spectronaut features include spectral library generation of spectral library from MaxQuant and Proteome Discoverer search results, fully automated in-run calibration (robust against mass shifts of up to 20 ppm), automatic quality control, peak scoring interference correction and intuitive data visualization.

9.4.4 Skyline

The open-source proteomics data analysis software package Skyline (www.sciex.com/products/software/skyline-software), developed by the

MacCoss Lab at the University of Washington, can be used for DIA and SWATH-MS data analysis. See chapter Proteomic Database Search and Analytical Quantification for Mass Spectrometry for further discussion of Skyline.

9.4.5 OpenSWATH

Another open-source proteomics data analysis software package is OpenSWATH (www.openswath. org), developed by the Aebersold Lab at the Institute for Molecular Systems Biology at ETH Zurich. This package can be used for generation of the spectral library, spectral alignment and targeted data extraction as described in Section 9.3, as well as statistical analyses.

9.5 Summary

SWATH-MS is a relatively novel mass-spectrometry-based approach that combines the strengths of DDA and DIA. DIA allows for precise quantification of proteins based on the area under the peak similar to multiple reaction monitoring (MRM) approach without the need for peptide labeling. Development of instrumentation, in particular the speed of data acquisition, has allowed for methods to be designed with narrow mass-to-charge windows, leading to the acquisition of subsets of data which are much more manageable. We recognize that if a protein is not represented by spectra in the library, it cannot be quantified by SWATH-MS. However, the requirement of generating a library of spectra as a prerequisite of SWATH-MS experiment is not a limiting factor because libraries of spectra can be extended by subsequent DDA acquisitions. Additionally, even as new versions of algorithms speed up data processing, as datasets are expanded (including libraries and SWATH acquisitions) more and more computer power will be needed. Even though we can have an intuitive sense of the nature of a small dataset, it is much harder to define a large dataset. In our view, current tools for statistical and bioinformatics analyses should be able to handle the ever-increasing datasets of SWATH-MS based experiments. We also think that there is much more to accomplish in proper experimental design and sample preparation relative to the analytical part of the overall proteomic study. Thus, the future development of SWATH-MS methodology will be focused on further understanding of sources of variability and how this can be reduced at various levels, ie, sample preparation, instrumental analysis and data processing.

References

[1] Legrain P, Aebersold R, Archakov A, Bairoch A, Bala K, Beretta L, et al. The human proteome project: current state and future direction. Mol Cell Proteomics 2011;10(7). http://dx.doi.org/10.1074/mcp.O111.009993.
[2] Snider J. Overview of mass spectrometry. Thermo Fischer Scientific Inc.; 2011 [Accessed].
[3] Egertson JD, Kuehn A, Merrihew GE, Bateman NW, MacLean BX, Ting YS, et al. Multiplexed MS/MS for improved data-independent acquisition. Nat Methods 2013;10:744—6.
[4] Geromanos SJ, Vissers JP, Silva JC, Dorschel CA, Li GZ, Gorenstein MV, et al. The detection, correlation, and comparison of peptide precursor and product ions from data independent LC-MS with data dependant LC-MS/MS. Proteomics 2009;9:1683—95.
[5] Xie H, Gilar M, Gebler JC. Characterization of protein impurities and site-specific modifications using peptide mapping with liquid chromatography and data independent acquisition mass spectrometry. Anal Chem 2009;81:5699—708.
[6] Bern M, Finney G, Hoopmann MR, Merrihew G, Toth MJ, MacCoss MJ. Deconvolution of mixture spectra from ion-trap data-independent-acquisition tandem mass spectrometry. Anal Chem 2010;82:833.
[7] Gillet LC, Navarro P, Tate S, Rost H, Selevsek N, Reiter L, et al. Targeted data extraction of the MS/MS spectra generated by data-independent acquisition: a new concept for consistent and accurate proteome analysis. Mol Cell Proteomics 2012;11:O111 016717.
[8] Law KP, Lim YP. Recent advances in mass spectrometry: data independent analysis and hyper reaction monitoring. Expert Rev Proteomics 2013;10:551—66.
[9] Collins BC, Gillet LC, Rosenberger G, Rost HL, Vichalkovski A, Gstaiger M, et al. Quantifying protein interaction dynamics by SWATH mass spectrometry: application to the 14-3-3 system. Nat Methods 2013;10:1246—53.

[10] Haverland NA, Fox HS, Ciborowski P. Quantitative proteomics by SWATH-MS reveals altered expression of nucleic acid binding and regulatory proteins in HIV-1-infected macrophages. J Proteome Res 2014;13:2109—19.

[11] Cheadle C, Vawter MP, Freed WJ, Becker KG. Analysis of microarray data using Z score transformation. J Mol Diagn 2003;5:73—81.

10

TOP-DOWN PROTEOMICS

C. Boone and J. Adamec

University of Nebraska – Lincoln, Lincoln, NE, United States

CHAPTER OUTLINE

10.1 Introduction   175
10.2 Protein Separation Methods   176
    10.2.1 Liquid Chromatography for Inline LC-MS Top-Down Proteomics   177
    10.2.2 Liquid Chromatography for Offline LC-MS Top-Down Proteomics   180
10.3 Mass Spectrometry of Intact Proteins   183
    10.3.1 Ionization Techniques in Top-Down Proteomics   184
    10.3.2 Mass Spectrometry Instruments for Top-Down Proteomics   185
10.4 Software for Data Analysis   188
References   190

10.1 Introduction

Traditionally, mass-spectrometry-based proteomics has been performed using the bottom-up approach, which involves the chemical or enzymatic digestion of proteins, peptide mass analysis, and inferred protein identification based on identified peptides. Limitations to this approach include protein inference problems and the inability to detect differing proteoforms [1]. The proteoform label refers to all possible molecular species of a protein product arising from a single gene. These include changes due to genetic variation, alternatively spliced transcripts, alternative translational start site, and posttranslational modifications [2]. The top-down

Proteomic Profiling and Analytical Chemistry. http://dx.doi.org/10.1016/B978-0-444-63688-1.00010-0

approach refers to the ionization of intact proteins and MS or MS/MS analysis of intact species or fragment ions generated in a mass spectrometer by induced dissociation [3]. Top-down proteomics can eliminate most of the problems associated with the bottom-up approach and allows for high-specificity protein identification and characterization of different proteoforms—information that would otherwise be lost with protein digestion and unattainable with other large-scale, whole-proteome approaches [4]. However, several issues in proteome-wide analysis of intact proteins have limited its use. A major obstacle in top-down proteomics is sample complexity. Biological samples are inherently complex, with a multitude of different proteins and their proteoforms; thus effective implementation of the top-down approach requires an extensive separation and/or enrichment step preceding mass analysis [5].

10.2 Protein Separation Methods

Protein separation minimizes ion suppression, increases the dynamic range of detection, and reduces precursor spectral complexity, simplifying data interpretation [6]. Techniques for separation of intact proteins are the same in principle as for the separation of peptides, but they are different in methodology. Sample separation can be directly coupled to the MS instrument (inline) or applied independent of the MS instrument (offline). Inline separation techniques have the advantage of reduced sample handling and increased throughput. However, time constraints due to continuously eluting molecules limit data collection and possibly the use of multiple fragmentation techniques [5]. Using offline separation involves fraction collection followed by direct infusion into the mass spectrometer, providing more time for data collection on a single fraction and enabling the use of multiple fragmentation techniques. Moreover, offline separation conditions do not need to be MS compatible, as fraction cleanup prior to infusion is possible; thus offline separations conditions are flexible.

10.2.1 Liquid Chromatography for Inline LC-MS Top-Down Proteomics

For the separation of intact proteins, liquid chromatography (LC) is routinely used [7]. Fundamentally, LC separation depends on the distribution of the proteins between the liquid mobile-phase solvent system, in which the proteins are initially contained, and the stationary phase. In mass spectrometric analysis of biological molecules, LC is often coupled with electrospray ionization; therefore LC is an efficient and valuable method for in-line analysis [8]. However, a critical point in the inline LC-MS analysis is the compatibility of solvents with MS analysis. Generally only three LC techniques are considered as LC-MS compatible: reversed-phase liquid chromatography (RPLC), hydrophilic interaction liquid chromatography (HILIC) and size-exclusion chromatography (SEC).

10.2.1.1 Reversed-Phase Liquid Chromatography

Reversed-phase liquid chromatography separates molecules based on surface hydrophobicity and is the most commonly used and widely applicable LC technique. RPLC is most commonly applied as the final dimension of separation. This is due to the ability of RPLC to exchange the original solvent for the MS-compatible solvent. RPLC uses a hydrophobic or nonpolar stationary phase and a hydrophilic or polar mobile phase. The stationary phase is commonly composed of porous silica particles linked to alkyl chains (C4, C5, C8, C18) or other inert nonpolar substances such as divinylbenzene (DVB). Shorter alkyl chains (C4 and C8) are typically preferred for intact protein separation because they are less retentive than longer alkyl chains [5]. Longer chains such as C18 can be used in special cases, typically for the small or low hydrophobic proteins (≤ 10 kDa). Larger proteins are usually much more hydrophobic and therefore strongly interact with the C18 matrix. This results in very broad chromatographic peaks during the elution and in many cases to incomplete elution. Nonporous silica (NPS) particles with fused alkyl chains may also be used in the stationary phase. Use of NPS in the stationary phase poses the advantage of increased protein recovery, but is limited by loading capacity [9]. As the sample is introduced to the stationary phase, the hydrophobic molecules in the polar mobile phase adsorb to the hydrophobic stationary phase, permitting the more hydrophilic molecules to be eluted first. Decreasing the polarity of the mobile phase by increasing the percent of organic solvent, usually acetonitrile, reduces hydrophobic interaction between the stationary phase and solutes, allowing for solute elution. More hydrophobic solutes will bind more strongly to the stationary phase, and thus a higher concentration of organic solvent is required in the mobile phase for their elution [5].

A typical gradient consist of two solvents including a polar solvent (0.1% formic acid) and an organic solvent (100% ACN; 0.1% formic acid). Following sample injection (proteins can be in MS noncompatible buffer with high salt concentration), proteins are retained on the column and potentially interfering salts are washed out with low organic solvent (3–5%). Proteins are then eluted in a gradient from 5% to 60% organic solvent in 30–90 min depending on the sample complexity. Two additional steps are also important: removal of any residual proteins from the column in the cleaning step achieved by high organic solvent for at least 2 min (100% ACN; 0.1% formic acid), and column reequilibration with polar solvent (0.1% formic acid) for at least 10 min (solvent exchange in the column can be monitored by back pressure: at high organic levels, pressure will drop 50% of original value and will get back after solvent is completely replaced). Another important aspect is column dimensions and pore sizes. The column should be at 10 cm long and the internal diameter (ID) can range from 75 μm (nano spray with flow rates 200–500 nl/min) up to 4.2 mm (flow rates 0.5–1.0 ml/min). Smaller ID increases sensitivity due to more efficient droplet formation; however, it is more difficult to operate and much less robust. Good compromise between sensitivity and robustness is usually achieved by columns with 0.5–1.0 mm ID and flow rates of 2–10 μl/min. The size of the pores in the stationary phase depends on the size of proteins to be analyzed. They can range from 300 Å for small proteins such us insulin up to 3000 Å for large molecules exceeding 150 kDa.

10.2.1.2 Hydrophilic Interaction Liquid Chromatography

In contrast to RPLC and HIC, hydrophilic interaction liquid chromatography (HILIC) uses a hydrophilic or polar stationary phase and a mobile phase with high organic content and gradient increasing in polar content; thus more hydrophobic species will elute first. The combination of the polar stationary phase and relatively nonpolar mobile phase creates a water-- enriched region surrounding the stationary phase. Proteins transition between this water-enriched region and the mobile phase. In comparison, in traditional normal-phase chromatography, analytes adsorb to the hydrophilic stationary phase [10]. Selection of the column size, flow rates and pore size is the same as for RPLC. A typical gradient consists of two solvents including organic solvent (ACN or methanol) and polar solvent (water). The sample is injected in high organic solvent (typically 80% ACN or methanol; 0.1% formic acid) and retained proteins are eluted in a gradient from 20% to 100% polar solvent (0.1% formic acid) in 30–90 min depending on the sample complexity. Very important is column reequilibration with organic solvent for at least 25 min.

10.2.1.3 Size-Exclusion Chromatography

Size-exclusion chromatography (SEC) was one of the first liquid chromatographic techniques developed and represents an excellent choice for protein–protein interaction analysis. As the name implies, SEC enables separation of molecules based on molecular weight or size. The stationary phase consists of a porous material. Small proteins are able to enter the pores, whereas larger proteins are unable to enter the pores. Therefore large molecules pass through the column faster and elute first; while smaller molecules get "trapped" within the particle pores, traverse a longer distance through the pores and elute toward the end of the chromatogram. The composition of the mobile phase is a very important factor for inline techniques. The solvent has to be compatible with MS analysis and prevent unspecific interactions with the stationary phase. Other issues include protein solubility, protein integrity and protein complex association, the latter two of which would be required for native SEC. Therefore ammonium bicarbonate buffer is usually used at concentration of 50–150 mM. SEC is a notoriously low-resolution separation technique. With the addition of ultra-high pressure, the resolution has greatly increased, being able to separate intact proteins from 6 to 670 kDa [11]. Flow rate, sample volume, column length, and particle pore size are main factors in chromatographic resolution of SEC. A higher flow rate results in higher resolution and sharper chromatographic peaks due to suppression of protein diffusion. On the other hand, high flow rate will result in incomplete separation, causing near simultaneous elution of proteins with differing size and creating spectral overlap between peaks in the chromatogram. Similarly, larger column length requires the sample to travel through more particles, permitting greater separation. Finally, larger pore size will permit larger proteins to enter, resulting in a longer elution time and higher resolution of larger species compared to a smaller pore size (Fig. 10.1). Optimal chromatographic parameters vary and strongly depend on the supplier, and therefore it is essential to follow all recommendations and suggestions supplied with the SEC column.

10.2.2 Liquid Chromatography for Offline LC-MS Top-Down Proteomics

As mentioned above, offline separation methods are not compatible with MS analysis and require either fraction collection and buffer exchange followed by direct infusion into the MS instrument or coupling with other inline-compatible methods in multidimensional separation platforms. This is due to a high concentration of nonvolatile salts such as NaCl in buffers typically used in the offline

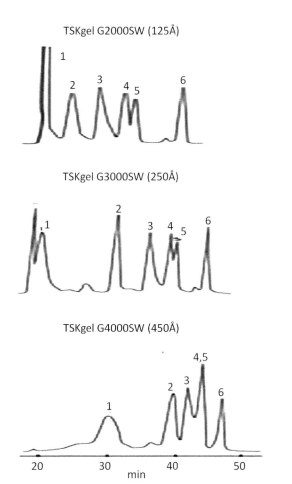

TSKgel G2000SW (125Å)

TSKgel G3000SW (250Å)

TSKgel G4000SW (450Å)

Figure 10.1 Effect of pore size on the elution of the intact proteins in SEC. With increasing pore size, the system can separate higher-molecular-weight proteins or protein complexes. However, peaks are broader and the low-molecular-weight proteins are not well resolved. (Sample: 1. Thyroglobulin. 2. Bovine serum albumin. 3. Beta-lactoglobulin. 4. Myoglobin. 5. Cytochrome C. 6. Glycine tetramer.) Adapted from TOSO Bioscience website, http://www. separations.us.tosohbioscience.com/.

separation techniques. Nonvolatile salts strongly interfere with ionization process, and may accumulate in the ion source and ion transfer system of the mass spectrometer, affecting its performance. Buffer exchange in collected fractions can be accomplished using commercially available spin columns that are based either on the reversed phase (C4) interaction or molecular filter with specific molecular weight cutoff (typically 5 kDa). Offline separation advantages, on the other hand, include more time for data collection, allowing application of multiple fragmentation techniques and optimize MS or MS/MS conditions and the use of the buffers favorable to protein conformation or protein–protein interactions.

10.2.2.1 Hydrophobic Interaction Chromatography

In theory, hydrophobic interaction chromatography (HIC) and RPLC are closely related, as in both techniques separation is based on hydrophobic interactions between the surface of an analyte and the stationary phase. However, in application the techniques are very different. The solid phase used in RPLC is characteristically more hydrophobic than that used in HIC. Therefore, RPLC results in stronger interactions between solute and solid phase compared to HIC. For elution from RPLC, organic solvents must be used. In comparison, the weaker hydrophobic interactions present using HIC can be disrupted by decreasing the concentration of salt in the mobile phase. HIC offers an alternative system to exploit hydrophobic properties of molecules in a more polar and less denaturing environment [12]. The HIC stationary phase consists of a nonionic group (octyl-, butyl-, hexyl-, phenyl-, propyl-) fused to an inert matrix, such as cross-linked agarose or sepharose. The mobile phase consists of a phosphate buffer, pH 7 and a salt such as potassium chloride, ammonium sulfate, or ammonium tartrate. A mobile phase containing a stronger salt, such as ammonium sulfate versus potassium chloride, causes a greater degree of protein denaturation, resulting in a greater degree of hydrophobic binding [12].

10.2.2.2 Ion-Exchange Chromatography

While fractionation using RPLC, HIC, and HILIC depend primarily on differing hydrophobicity, ion-exchange chromatography (IEX) achieves separation due to differences in analyte charge, which strongly depends upon mobile-phase pH. The stationary phase used in IEX is composed of a matrix, usually porous beads that are composed of cross-linked polysaccharides, synthetic organic polymers, or inorganic materials, and an immobilized ligand, either positively or negatively charged. Positively charged immobilized ligands are called anion exchangers and are utilized in either strong or weak anion exchange chromatography (SAX, WAX). Negatively charged immobilized ligands are called cation exchangers and are utilized in either strong or weak cation exchange chromatography (SCX, WCX). Intact proteins in a mobile phase with a pH less than their pI will be positively charged and thus bind to cation exchangers. Conversely, intact proteins in a mobile phase with a pH greater than their pI will be negatively charged and thus bind to anion exchangers. Therefore, pH of the mobile phase is an essential factor in IEX (typical pH of mobile phases for AX and CX are 8.5 and 6.5, respectively). The mobile phase starts with low salt concentrations ($\sim$25 mM), which allows for protein binding and increasing the salt concentration (up to 1.5 M) causes protein elution. Also, for weak ion exchangers, changes in pH may result in changes in charge, and therefore applying a pH gradient in weak ion exchange may cause elution. This variant of IEX is called chromatofocusing [13].

10.3 Mass Spectrometry of Intact Proteins

Although MS analysis of intact proteins has existed for decades, the identification and detailed analysis of intact proteins including the structural, functional, and dynamic characterization was enabled by implementation of soft ionization techniques together with development of high-resolution/accuracy instruments. Progress to stabilize proteins and protein complexes during ionization has shown great promise in native electrospray MS for protein complex identification up to as large as 1 MDa [14]. Similarly, supercharging ionization coupled with native electrospray mass spectrometry has been successfully used to identify metal or nucleotide ligand binding sites within approximately 20 amino acids [15].

10.3.1 Ionization Techniques in Top-Down Proteomics

MS techniques can detect ionized molecules only; therefore proteins must acquire a charge before analysis. The ionization process must be highly efficient but at same time limits fragmentation of the proteins. Among many ionization techniques, matrix-assisted laser desorption/ionization (MALDI) [16] and electrospray ionization (ESI) [17] are almost exclusively used for the ionization of the intact proteins. While the MALDI technique for ionization can be used for offline top-down proteomics, ESI is more universal and can be applied to both inline as well as offline strategies.

Prior to MALDI-MS analysis, proteins contained in individual fractions must be desalted. This is typically done using commercially available C4-based ZipTips or spin columns. Fraction is loaded into ZipTip by pipetting, salts washed out using polar solvent (0.1% formic acid) and proteins eluted in a low volume of high organic solvent (80% ACN; 0.1 formic acid). At this point, proteins must be immediately mixed with matrix solution (1:1 v/v) and 0.5–2 µl is deposited directly on the MALDI plate to prevent protein precipitation in high organic solvent and allowed to dry. Very important is the selection of the matrix—typical matrices are α-cyano-4-hydroxycinnamic acid (CCA) for small proteins, 3,5-dimethoxy-4-hydroxycinnamic acid (sinapinic acid or SA) for large proteins, and 2,5-dihydroxybenzoic acid (DHB) for glycoproteins. All matrices must be freshly prepared in 50% ACN and 0.1% trifluoroacetic acid (TFA). The dried MALDI spot containing sample and matrix is then pulsed with a laser and the energetic matrix transfers charge to the protein [16], which is detected by MS. Another option is to dry cleaned samples in SpeedVac, resolubilize them in 0.1% formic acid (5% ACN can be added to help the solubilization process) and keep them at −80°C until further analysis. The great advantage of the MALDI-MS is that it generally produces singly charged proteins (some double- or triple-charged molecules can be also observed depending on amino acid composition and protein size). Therefore data interpretation is simple and observed molecular mass directly corresponds to the single charged mass of the protein.

The ESI, on the other hand, deposits positive charges to accessible basic sites of proteins during the ionization process. Positive charge is typically added through proton, sodium, potassium and ammonium adducts. High-molecular-weight species, such as intact proteins, contain a large number of elements composing the amino acid building blocks. This inherent feature of intact proteins attracts multiple positive charges and produces a broad range of multiple charge state peaks (Fig. 10.2B). In addition, each charge state consists of multiple peaks reflecting isotope distribution (Fig. 10.2C). Collectively, these two effects generate a complex spectral signal from a single proteoform that may easily be spread across a wide mass-to-charge ratio (m/z) and reducing the sensitivity of the analysis [18]. Additionally, these effects will create a complex precursor spectra from a sample containing multiple proteoforms or fragmentation spectra with a large number of fragment ions. Therefore, to separate and correctly assign peaks arising from these complex spectra, high resolution and mass accuracy are essential [19]. High resolution is required to distinguish proteoforms containing phosphorylation versus sulfation ($\Delta m = 10$ mDa), trimethylation versus acetylation ($\Delta m = 39$ mDa), deamidation ($\Delta m = 1$ Da), or disulfide bond formation ($\Delta m = 2$ Da) [20]. Example of such structural analysis, data interpretation identifying a single disulfide bond and cysteine oxidation are shown in Fig. 10.3.

10.3.2 Mass Spectrometry Instruments for Top-Down Proteomics

As mentioned earlier, top-down proteomics requires high-resolution/accuracy MS instruments. Of those, the Fourier Transform Ion Cyclotron Resonance MS (FT-ICR-MS), commercially available through Bruker and Thermo, and the FT-Orbitrap-MS available through Thermo, are instruments fully capable of detailed characterization due to their

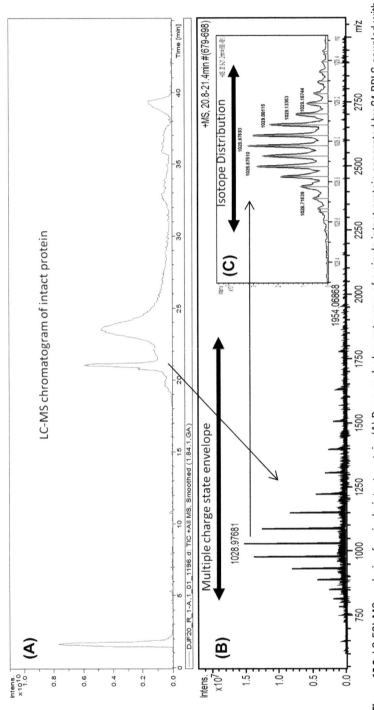

Figure 10.2 LC-ESI-MS analysis of a single intact protein. (A) Base peak chromatogram of a single intact protein separated by C4 RPLC coupled with Solarix 7T FT-ICR MS. (B) MS spectra of single intact protein indicating multiple charge states (each peak corresponds to the specific charge state). (C) Detailed MS spectra of the peak at 1028 m/z indicating isotope distribution inside the specific charge state.

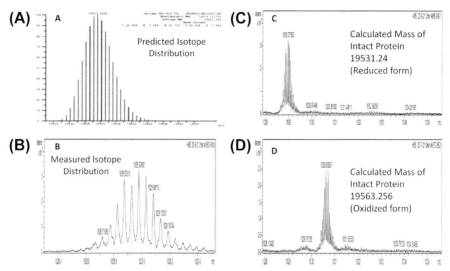

Figure 10.3 Data interpretation of PTMs in the intact protein LC-ESI-MS analysis. (A) Theoretically predicted isotope peaks distribution of the intact protein DJ-1. (B) Experimentally obtained isotope peaks distribution of DJ-1 protein. Inconsistency in the peak distribution suggests two forms of DJ-1. Peak fitting and modeling revealed that ~10% of DJ-1 is intraconnected through a disulfide bond. (C) Reduced form of the DJ-1 protein with calculated molecular mass 19,531.24 Da. (D) Oxidized form of DJ-1 protein with calculated molecular mass 19,563.56 Da. The difference, 32.32 Da, represents either sulfinic modification of the single cysteine or sulfenic modification of two cysteines.

resolving power. While FT-MS instruments demonstrate superior resolving power and mass accuracy necessary for top-down proteomics, realistically these instruments are much more expensive and require a high level of expertise. Therefore time-of-flight instruments still represent an acceptable alternative. Specific to FT-ICR-MS instruments is a superconducting magnet and analyzer cell (ICR cell). The superconducting magnet functions in producing a magnetic field that traps charged particles in the radial direction on a circular trajectory. The greater the strength of the magnet, the greater the magnetic field produced and the greater the resolving power of the instrument [21]. FT-ICR-MS instruments have a high resolving power with dependence on magnetic field strength of approximately 1,000,000 at $m/z = 400$ for a 12 T instrument and mass accuracy in the range of 1 to 0.05 ppm. Dynamic range with an upper limit of approximately 2000 m/z.

The Orbitrap MS is composed of a barrel-like outer electrode and a spindle-like central electrode. Stable ion trajectories of orbiting motion around the central electrode and simultaneous oscillations in the z-direction result from applying DC voltage between the two axially symmetric electrodes [22]. Similar to FT-ICR instruments, Orbitrap instruments are coupled to linear ion traps or linear trap quadrupoles (LTQ) and utilize Fourier transformation in spectra generation [21]. An important difference between FT-ICR and Orbitrap instruments, especially when analyzing intact proteins, is the decrease in instrument resolving power at larger m/z values. The electrostatic field in an Orbitrap mass analyzer causes a much slower drop in resolving power, compared to an ICR instrument for ions of increased m/z. Therefore, the Orbitrap mass analyzer may theoretically outperform the FT-ICR mass analyzer for ions above a certain m/z. Orbitrap mass analyzers have a high resolving power, generally exceeding 120,000; mass accuracy in the range of 2–5 ppm; and an m/z range upper limit of at least 5000.

Time-of-flight (TOF) instruments generally do not display the high mass resolving power or mass accuracy in comparison to FT mass spectrometers (ICR and Oribitrap). TOF instruments equipped with a reflectron have a mass resolving power of 10,000 to 20,000 at m/z of 400 and mass accuracy between 2 and 30 ppm. However, the dynamic mass range of a TOF instrument is theoretically unlimited, dependent on the length of the flight or field free drift tube. Development of the Q-TOF-MS instruments has dramatically increased the mass resolving power up to approximately 50,000 at m/z of 400, while still maintaining a large dynamic range, up to 1.5 MDa for some instruments.

10.4 Software for Data Analysis

The Department of Computer Science and Engineering, University of California in San Diego, provides two key software tools for top-down MS: MS-Deconv (http://bix.ucsd.edu/projects/msdeconv)

and MS-Align+ (http://bix.ucsd.edu/projects/msalign). Deconvolution in top-down MS is critically important due to multiple charge states and isotopomer envelope grouping. MS-Deconv uses sets of charge-state envelopes to determine the spectral graph rather examining charge-state envelopes individually. MS-Align+ is a spectral alignment tool that allows for searches for unexpected PTMs. Another software package commercially available from Thermo Fisher is ProSightPC 3.0 [23]. Currently, ProSightPC 3.0 adequately supports high mass accuracy tandem MS experiments performed on the FT-ICR and the Orbitrap, including the Q Extractive. It automatically detects and annotates PTMs in database files that are in the UniProtKB flat file format. The software also creates a proteome warehouse that stores data on proteome forms and fragment ions from proteome forms based on different databases including UniProtKB flat and MySQL. ProSightPC 3.0 also allows the user to create his or her own database with FASTA-formatted text files. A variety of searches may be used, including absolute mass, biomarker, sequence tag, single protein, gene-restricted absolute mass and gene-restricted biomarker in ProSightPC 3.0 (Thermo Scientific ProSightPC 3.0 user guide).

Regardless of the software, there are several important parameters that have to be considered in order to successfully deconvolute and interpret the intact protein MS data. Every MS instrument generates background level peaks of low-level intensity that represent noise. To distinguish between real peaks corresponding to the specific molecule and noise, the signal to noise ratio (S/N) is defined. Typically this parameter is set to 3. Lower values can be used too; however, this will increase the number of false-positive identifications. Two sets of mass ranges must be also defined. While scan mass range represents a range of m/z acquired by instrument (typically 250–4000 m/z), minimum and maximum mass of expected proteins defines a range of molecular masses of the proteins of interest (in complex mixtures it may range from 5 to 250 kDa; if the molecular mass of expected proteins is known than the corresponding range should be used). A very important parameter is the maximum charge state. As described earlier, the intact proteins attract multiple positive charges and produce a broad range of multiple charge states peaks (Fig. 10.2B). The larger a protein is, the more charges it attracts. Even high-molecular-weight proteins can be detected in a 250–4000 m/z range with as many as 100 charges. Therefore the parameter should be set relatively high for the complex biological samples (eg, 100 or 150). Precursor/fragment errors are defined by instrument accuracy and typically ranging between 5 and 20 ppm.

References

[1] Nesvizhskii AI, Aebersold R. Interpretation of shotgun proteomic data: the protein inference problem. Mol Cell Proteomics 2005;4:1419–40.

[2] Smith LM, Kelleher NL. Proteoform: a single term describing protein complexity. Nat Methods 2013;10:186–7.

[3] Garcia B. What does the future hold for top down mass spectrometry? J Am Soc Mass Spectrom 2010;21:193–202.

[4] Scheffler K. In: Martins-de-Souza D, editor. Top-down proteomics by means of orbitrap mass spectrometry. New York: Springer; 2014. p. 465–87.

[5] Capriotti AL, Cavaliere C, Foglia P, Samperi R, Laganà A. Intact protein separation by chromatographic and/or electrophoretic techniques for top-down proteomics. J Chromatogr A 2011;1218:8760–76.

[6] Lu P, Vogel C, Wang R, Yao X, Marcotte EM. Absolute protein expression profiling estimates the relative contributions of transcriptional and translational regulation. Nat Biotech 2007;25:117–24.

[7] Xie F, Smith RD, Shen Y. Advanced proteomic liquid chromatography. J Chromatogr A 2012;1261:78–90.

[8] Wu Q, Yuan H, Zhang L, Zhang Y. Recent advances on multidimensional liquid chromatography–mass spectrometry for proteomics: from qualitative to quantitative analysis—a review. Anal Chim Acta 2012;731:1–10.

[9] Meng F, Cargile BJ, Patrie SM, Johnson JR, McLoughlin SM, Kelleher NL. Processing complex mixtures of intact proteins for direct analysis by mass spectrometry. Anal Chem 2002;74:2923–9.

[10] Buszewski B, Noga S. Hydrophilic interaction liquid chromatography (HILIC)—a powerful separation technique. Anal Bioanal Chem 2012;402:231–47.

[11] Chen X, Ge Y. Ultra-high pressure fast size exclusion chromatography for top-down proteomics. Proteomics 2013;13. http://dx.doi.org/10.1002/pmic.201200594.

[12] Cummins P, O'Connor B. In: Walls D, Loughran ST, editors. Hydrophobic interaction chromatography. Humana Press; 2011. p. 431−7.

[13] Jungbauer A, Hahn R. In: Richard RB, Murray PD, editors. Chapter 22 Ion-exchange chromatography. Academic Press; 2009. p. 349−71.

[14] Clarke DJ, Campopiano DJ. Desalting large protein complexes during native electrospray mass spectrometry by addition of amino acids to the working solution. Analyst 2015;140:2679−86.

[15] Yin S, Loo JA. Top-down mass spectrometry of supercharged native protein−ligand complexes. Int J Mass Spectrom 2011;300:118−22.

[16] Karas M, Hillenkamp F. Laser desorption ionization of proteins with molecular masses exceeding 10,000 daltons. Anal Chem 1988;60:2299−301.

[17] Fenn JB, Mann M, Meng CK, Wong SF, Whitehouse CM. Electrospray ionization for mass spectrometry of large biomolecules. Science 1989;246:64−71.

[18] Compton PD, Zamdborg L, Thomas PM, Kelleher NL. On the scalability and requirements of whole protein mass spectrometry. Anal Chem 2011;83:6868−74.

[19] Breuker K, Jin M, Han X, Jiang H, McLafferty FW. Top-down identification and characterization of biomolecules by mass spectrometry. J Am Soc Mass Spectrom 2008;19:1045−53.

[20] Zhang K, Yau PM, Chandrasekhar B, New R, Kondrat R, Imai BS, et al. Differentiation between peptides containing acetylated or tri-methylated lysines by mass spectrometry: an application for determining lysine 9 acetylation and methylation of histone H3. Proteomics 2004;4:1−10.

[21] Scigelova M, Hornshaw M, Giannakopulos A, Makarov A. Fourier transform mass spectrometry. Mol Cell Proteomics 2011;10. http://dx.doi.org/10.1074/mcp.M111.009431−19.

[22] Makarov A. Electrostatic axially harmonic orbital trapping: a high-performance technique of mass analysis. Anal Chem 2000;15:1156−62.

[23] Dang X, Scotcher J, Wu S, Chu RK, Tolić N, Ntai I, et al. The first pilot project of the consortium for top-down proteomics: a status report. Proteomics 2014;14:1130−40.

11

PROTEOMIC DATABASE SEARCH AND ANALYTICAL QUANTIFICATION FOR MASS SPECTROMETRY

M. Wojtkiewicz, J. Wiederin and P. Ciborowski

University of Nebraska Medical Center, Omaha, NE, United States

CHAPTER OUTLINE

11.1 Introduction   194
11.2 Protein Databases   196
    11.2.1 SwissProt   197
    11.2.2 UniProt   197
    11.2.3 UniRef   197
    11.2.4 National Center for Biotechnology Information Nonredundant Database   198
    11.2.5 Other Databases and Resources   198
11.3 Search Engines   199
    11.3.1 Mascot   200
    11.3.2 SEQUEST and Proteome Discoverer   201
    11.3.3 Protein Pilot   202
    11.3.4 MaxQuant   202
    11.3.5 X! Tandem or the Global Proteome Machine   203
    11.3.6 SpectraST   203
11.4 Mass Spectrometry Data Searches: Things to Consider   203
    11.4.1 Search Parameters—Mass Tolerance   204
    11.4.2 Miscleavage: Friend or Foe?   205
11.5 Post-Database Search Data Processing   205
    11.5.1 Trans Proteomics Pipeline   206
    11.5.2 Scaffold   206

Proteomic Profiling and Analytical Chemistry. http://dx.doi.org/10.1016/B978-0-444-63688-1.00011-2

11.5.3 ProteoIQ   207
11.5.4 Skyline   207
11.6 Searches for Posttranslational Modifications   207
11.7 Summary   208
References   209

11.1 Introduction

As the field of proteomics evolves and the operation of mass spectrometers becomes more user-friendly, an increasing number of chemists and biologists are becoming routinely involved in designing and executing proteomic studies, reflected by the increasing number of publications. At the same time, proteomic datasets are becoming larger and more complex, requiring not only more computer power for processing but also more in-depth understanding of the tools implemented in database searches and the calculation of statistical significance even before data are subjected to bioinformatics analyses. Technological development and the complexity of proteomic profiling studies has made manual interrogation of spectra almost inappropriate and created an urgent demand for computerized (automated) tools for tandem mass spectra interpretation. It is important to note that as more comprehensive pipelines for proteomic data analysis are available, in this chapter we focus on tools and procedures that have been vetted in two decades of collective experience of all scientists contributing to development of the proteomics field. It is not exhaustive, due to ever-changing expansions related to databases, databases searches and post-database data processing.

Peptide sequencing by tandem mass spectrometry contains two important pieces of information: the mass of the precursor ion (MS) and the masses of fragments (MS/MS). This information is matched against comprehensive protein sequence databases using any number of available search engines. Search engines utilize different algorithms to match theoretical peptide fragmentation with corresponding precursors and fragments measured during mass spectrometric analysis. Indexing peptide entries based on the specific method of protein digestion is highly beneficial for matching fragments because it significantly reduces time during database searching. For example, we know that trypsin cuts peptide bonds at the C-terminal position of lysine and arginine. Indexing the database for the specific enzyme will reduce the number of possible peptides derived from proteins. It is also important to recognize that the C-terminal of a peptide or protein will not necessarily contain lysine or arginine.

With very few exceptions, software for database searches gives the investigator power to set search parameters. Mass accuracy of the precursor ion plays a very important role because many peptides may have very similar molecular weight as a combination of various amino acid sequences. Therefore, manufacturers have developed mass spectrometers to meet the requirements of high mass accuracy as well as high resolution. Orbitrap technology combined with an ion trap serves as an excellent example. This particular setting measures masses of precursor ions in the orbitrap for a prolonged period of time (high resolution), while fragmenting and analyzing selected precursors in the ion trap (high sensitivity). Other methods can be based on platforms such as quadrupole-time-of-flight (qTOF). A list of precursor ions based on their intensity is prioritized to decide which are fragmented. After analysis, the instrument software can generate a peak list, including the precursor ion mass-to-charge (m/z) values, as well as charge states and a list of their fragment ions. Once this peak list is created and if needed, converted into another format, the searching and scoring process proceeds. The output is dependent on the search engine.

This chapter describes database searches and processing for data-dependent acquisition. As data independent acquisition is an expanding part of proteomics, we have dedicated a separate chapter to SWATH-MS (chapter SWATH-MS: Data Acquisition and Analysis). In brief, this chapter aims to provide the "working knowledge" necessary for research scientists, who are not proteomics experts, to understand approaches to processing and analysis of proteomic data.

11.2 Protein Databases

The quality of information extracted from mass spectrometry data depends not only on the quality of algorithm(s) used for database searches, but also on the quality and accuracy of the databases themselves. After the analytical phase of proteomic profiling is completed and all MS and MS/MS spectra are transformed into a searchable format, choosing the database and extraction of information is the crucial next step. There are many databases available, but a "nonredundant" database is preferred. Redundancy in a theoretical protein database can occur in many ways. An example is whether two alleles of the same locus or two isoenzymes from the same organism constitute redundancy, and where the borderline is between redundancy and completeness of information included in database. No answer is simple and the high complexity of biological data makes it impossible to apply a generic definition of redundancy. For example, the curated UniProt database provides a canonical sequence and other related sequences of isoforms and variants to address this issue.

The following section discusses the rationale for choosing a particular database to complete a search. The choice of database is an important factor that depends on the purpose of the project, which determines the amount of search space and the ease of data compilation. If the purpose is to characterize a reasonable number of proteins across a very large biological sample pool and throughput is a major concern, then a well-annotated, nonredundant database is a good choice. If the purpose is to identify all the sequence polymorphisms in a small number of noncomplex biological samples, then a large database such as the NCBI nr protein database is ideal. Specifying taxonomy parameters allows the user to conduct a search within the database for a particular species (or a group of closely related species). This will decrease the unnecessary search space and increase throughput.

11.2.1 SwissProt

Swiss-Prot, created in 1986, is a biological database of protein sequences that is manually curated, or reviewed and edited by experts. For that reason, Swiss-Prot provides reliable protein sequences associated with annotation with a minimal level of redundancy. In 2002, collaboration between the Swiss Institute of Bioinformatics, the European Bioinfomatics Institute and the Protein Information Resource (PIR), funded by the National Institutes of Health, formed the UniProt consortium, combining Swiss-Prot and its automatically curated supplement TrEMBL (the Protein Information Resource protein database), creating the most comprehensive catalog of protein information. An updated UniProtKB/Swiss-Prot protein knowledgebase release and statistics can be found at http://web.expasy.org/docs/relnotes/relstat.html.

11.2.2 UniProt

The Universal Protein Resource (UniProt) is a collaborative effort of three institutions: the European Bioinformatics Institute (EBI), the Swiss Institute of Bioinformatics (SIB) and the Protein Information Resource (PIR). The joint effort of UniProt [1] is to provide a comprehensive, high-quality and freely accessible resource of protein sequence and functional information. This is accomplished through various avenues, which include database-curating, development of nonlicensed software and data annotation. The UniProt database includes the UniProt Knowledgebase (UniProtKB), the UniProt Reference Clusters (UniRef), and the UniProt Archive (UniParc). The UniProt Metagenomic and Environmental Sequences (UniMES) database is a repository specifically developed for metagenomic and environmental data. Moreover, the UniProt database provides links to multiple related resources.

11.2.3 UniRef

The Universal Protein Resource Consortium [2,3] released UniRef in 2004 as a set of curated comprehensive databases. Under UniRef there are three separate sequence clusters constructed by merging sequences that have 100% (UniRef100), ≥90% (UniRef90) or ≥50% (UniRef50) sequence identity, regardless of source organism. The version(s) of these clusters should be included in reporting proteomic data.

11.2.4 National Center for Biotechnology Information Nonredundant Database

The National Center for Biotechnology Information (NCBI) formed in 1988 as a division of the National Library of Medicine (NLM) at the National Institutes of Health (NIH). Among other responsibilities, the NCBI facilitates the use of databases and software and performs research on advanced methods of computer-based information processing for analyzing the structure and function of biologically important molecules including proteins.

The "nr" database is the largest database available through NCBI BLAST. The name "nr" is derived from "nonredundant," but this is historical only, because this database is no longer nonredundant. This database is compiled by NCBI as a database for BLAST search and it contains nonidentical sequences from GenBank CDS (coding sequence) translation, Protein Data Bank (PDB), Swiss-Prot, Protein Information Resource (PIR) and Protein Research Foundation (PRF). The NCBI database is not updated at a fixed time interval; therefore it is important to note that protein sequence databases are in constant flux. With each round of database release, some protein sequences disappear; the annotation and accession numbers of the remaining sequences could change with the addition of new sequences. It is necessary to report the version of the database used for the searching sequences for reproducing a study by either one's own laboratory or the greater research community. For more information and downloads we recommend http://www.ncbi.nlm.nih.gov/

11.2.5 Other Databases and Resources

The International Protein Index (IPI), compiled and maintained by the European Bioinformatics

Institute (EBI), aims to provide a top-level guide to the main databases that describe the proteomes of higher eukaryotic organisms. Due to the similar function it served as compared to UniProt knowledge hub, the effort of maintaining IPI ceased in September 2011. Previously released versions of IPI databases are still available through the former IPI website.

The Human Genome and Protein Database (HGPD; http://www.HGPD.jp/) launched in 2008 was created to provide an online resource for the functional and structural analysis of gene products, ie, proteins. The HGPD 33,275 human Gateway entry clones have been constructed from the open reading frames (ORFs) of full-length cDNA, as well as other sequences deposited in public databases, such as RefSeq, Ensemble, Human ESTs, etc., thus representing the largest collection in the world [4].

PhosphoSitePlus (PSP) (www.phosphosite.org) is an online resource for protein modifications, such as phosphorylation, acetylation, ubiquitination and methylation. The National Institutes of Health (NIH), National Cancer Institute (NCI), National Institute on Alcohol Abuse and Alcoholism (NIAAA) and National Institute of General Medical Sciences (NIGMS) provide grant funding to Cell Signaling Technology (CST) for maintaining the website. The major goal of PSP is to provide a comprehensive systems biology website where researchers can retrieve information and have tools to study posttranslational modifications in the context of biological regulation, subcellular location, disease, etc. PSP is routinely modified and updated to stay relevant with rapidly evolving proteomic technology and data [5].

11.3 Search Engines

As presented in Fig. 11.1, a number of search engines are available to assess how likely (or unlikely) a particular mass list containing both precursor and fragment ions' masses represents a peptide sequence [6,7]. Each search engine uses a different algorithm and we expect each will have inherent strengths and weaknesses; therefore, it becomes obvious that

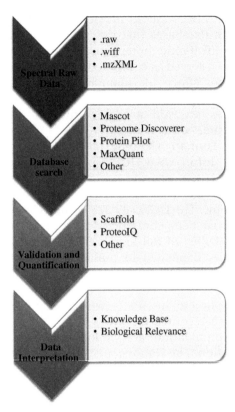

Figure 11.1 An overview of proteomic data processing using search engines.

these algorithms offer complementary approaches. Thus it is advisable to analyze each set of mass spectrometry data using more than one type of search engine to obtain a more comprehensive set of results. However, there are a couple caveats with using multiple search engines, which include a higher workload as well as the necessity of combining and narrowing search results.

11.3.1 Mascot

In the early 1990s, **M**olecular **W**eight **S**earch (MOWSE) was developed for protein identification by peptide mass fingerprinting [8]. The second-version release, MOWSE II, included amino acid sequence and composition qualifiers. MOWSE III soon followed and supported all the proven methods of protein identification: peptide-mass fingerprinting, MS/MS fragment ion search, and searches combining mass data with amino acid sequence or composition. Indexing MOWSE databases facilitated rapid and necessary calculations to search any FASTA-format database (http://www.matrixscience.com/help/history.html). In 1998, Matrix Science, Inc. commercialized MOWSE and changed the name to Mascot.

A major feature of Mascot's approach was to integrate multiple proven methods of searching [7,8] with a probability-based scoring system. Simply put, the scoring system reports a peptide ion score as $-10 \log (P)$, where P is the absolute probability that the observed match is a random event. Protein Mascot scores are derived from the sum of all the peptide scores belong to this protein. Mascot offers three components: (1) peptide mass fingerprint, which considers only peptide mass values as experimental data; (2) sequence query, which combines peptide mass data, amino acid sequence and composition information; and (3) MS/MS ion search, which uses uninterpreted MS/MS data from one or more peptides.

11.3.2 SEQUEST and Proteome Discoverer

The Sequest Algorithm, developed in the early 1990s in response to the demand for higher throughput, applies an approach that compares predicted mass spectra to empirical spectra, using a "closeness-to-fit" method in a two-step process. The first step calculates the preliminary score and the second step calculates the cross-correlation function to obtain a score (Xcorr) for the degree to which each candidate theoretical peptide matches the experimental query mass [6]. Therefore, the Sequest algorithm is based on a scoring system, which makes it different from probability-based approaches. The Sequest algorithm was the basis of BioWorks™ (ThermoScientific, Inc.) software package, which was further developed over the years and eventually transformed to the Proteome Discoverer™ package, with the intention of providing tools for database searches and analysis of custom-designed proteomic experiments. Proteome Discoverer is equipped with multiple features including Mascot search capability and support for multiple dissociation techniques (CID, ETD, and HCD). The ability to combine results facilitates identifying more proteins and PTMs. The software also supports isobaric mass tagging (TMT, iTRAQ), HeavyPeptide techniques for label-free relative or absolute quantitation, calculation of the false discovery rate (FDR), tools for validation of protein IDs, and automated annotation of identified proteins with GO classifications. Literature references from public databases illustrate biological context and have the ability for expansion through integration of custom software tools. These features allow researchers to perform individualized data analysis.

11.3.3 Protein Pilot

SCIEX (previously MDS Sciex or ABSciex) introduced the Paragon algorithm in 2007 [9], as part of the ProteinPilot Software package performing peptide identification. While able to interface with a Mascot server license, ProteinPilot uses the Paragon algorithm to automatically conduct protein interface analysis. One of the goals while developing the Paragon algorithm was to create a search engine with a limited set of manipulated search parameters that can be widely used by biologists, not only by computer programmers. There are two components of peptide identification in this algorithm: a sequence tag component and a standard precursor mass-filtered database search. The software's most recent version automatically exports into a spreadsheet format, but can also be converted into other formats.

11.3.4 MaxQuant

MaxQuant [10] is a free software package specifically tailored for analysis of high-resolution mass spectrometry (Orbitrap and FTMS) data. MaxQuant is mostly used for identification and quantification of SILAC-labeled, or TMT or iTRAQ-labeled samples, but can also be used in label-free identification and quantification. Raw files are processed for peptide identification through the Andromeda search algorithm [11] that calculates a probability score for peptide-spectrum matches. Companion software, known as Perseus, is where output files are subjected to statistical analysis and protein grouping.

11.3.5 X! Tandem or the Global Proteome Machine

A variety of search algorithms are available freely from the Global Proteome Machine. One of these, X! Tandem [12], is open-source software that uses an application programming interface (API). While simple, some programming knowledge is advisable. X! Tandem takes an XML file of instructions on its command line, and outputs the results into an XML file. It automatically calculates statistical confidence of both peptide and protein, which eliminates the need for secondary software.

11.3.6 SpectraST

Although the previously discussed search engines use theoretical protein databases to identify proteins, it is possible to use actual spectral data as a means of identification. SpectraST is short for "Spectra Search Tool" and is a spectral library-building and searching tool designed for bottom-up proteomics applications. It is now a component of the Trans Proteomic Pipeline (TPP) discussed later in this chapter.

11.4 Mass Spectrometry Data Searches: Things to Consider

After choosing a database, search engine(s) and knowing the instrument used for data acquisition, we need to consider several other parameters. We present below a brief description of basic necessary parameters for consideration in customizing search parameters: mass tolerance and miscleavages.

11.4.1 Search Parameters—Mass Tolerance

The two most common methods for protein and peptide ionization are electrospray ionization (ESI) and matrix-assisted laser desorption ionization (MALDI). Different types of instruments use different ionization methods, as well as different fragmentation types in the MS/MS stage [13]. MALDI tends to produce singly and doubly charged ions, while ESI tends to produce multiply charged (≥ 2) peptide ions during ionization. Different mass analyzers have different capacities of resolving power and mass accuracy that can be reproducibly achieved, ranging from high mass accuracy (Fourier transform ion cyclotron, Orbitrap) [14,15] to medium (time-of-flight) to low (ion trap, quadrupole) [16]. Depending on the instrument configuration, the categories of both parent ion and the fragment ions could be very different. Defining the correct instrument type for the spectrum data is critical to put the data on the right path for database searching.

Search algorithms allow users to set individual search parameters, including mass tolerance for precursor and fragment ions. The mass tolerance for both the precursor peptide ion and the fragment ions has a huge impact on the database search [17], as the tolerance determines the number of candidate peptides considered for matching with the query peptide. Appropriately defining the parameters of mass tolerance is essential for effectively harnessing the mass resolution of a particular instrument and reducing the search computation intensity. For example, the high mass accuracy and high resolving power of an Orbitrap or FTICR mass spectrometer would be wasted if one sets the mass tolerance to >10 ppm, producing false-positive results and larger protein lists for the investigator to muddle through. On the other hand, for a low-resolution ion trap mass spectrometer, one must loosen the mass tolerance to ~ 0.5 to 2.0 Da for consideration of a reasonable number of theoretical candidate peptides to match with the query.

11.4.2 Miscleavage: Friend or Foe?

Many enzymes and chemicals cleave peptide bonds for digestion of proteins. Here, we illustrate the issue of miscleavage with trypsin, the most commonly used enzyme. Trypsin cleaves peptide bond C-terminally to lysine and/or arginine and typically generates peptide fragments ranging from 10 to 20 amino acids long. This is a suitable length to more efficiently ionize (compared to the intact protein), while maintaining high enough sequence specificity. Nevertheless, tryptic digestion is not 100% complete, sometimes generating peptide fragments with an internal Lys and Arg residue, which represent miscleavages. Therefore, one must alter search parameters to include whether miscleavages are allowed under experimental conditions, and if so, then the number of miscleavages must be specified as 1 or 2. Specifying a higher number of miscleavages should be avoided unless there is a good reason (ie, studying peptide phosphorylation using electron transfer dissociation—in this case proteins are deliberately processed to be partially digested so that multiply charged peptides could be produced). Each additional missed cleavage site increases the number of calculated theoretical masses to be matched to the experimental data and therefore increases the computational overhead committed to searching.

11.5 Post-Database Search Data Processing

Algorithms for database searches provide researchers with high confidence identification of peptides and proteins along with additional parameters, ie, peak intensity, peak width at half height, and area under the peak for quantification of differences. Visualization of data is a subsequent step to help comprehend all the information contained in a usually large dataset. For quantification of differences in protein expression, results must go through the scrutiny of statistical analyses. This step can be performed either using component's built-in software packages or as standalone protocols using software for statistical analysis such as R, SSPS or Prism. Once statistically significant differences are revealed, collected information might be analyzed using bioinformatics tools. Additionally, all necessary information needed to set subsequent MRM validations can be derived from the mass of the precursor ion and transitions corresponding to such a precursor.

11.5.1 Trans Proteomics Pipeline

Discussed earlier, SpectraST and X! Tandem are available components of the open-source pipeline known as the Trans Proteomics Pipeline [18]. However, numerous other search engine results can be used in the pipeline as long as they are converted beforehand to .mzXML or .mzML. After database searching, the pipeline handles probability assignment and validation though several options such as PeptideProphet, which is based on the expectation maximization algorithm. The pipeline has options for quantification as well. Protein assignment is done through a component called ProteinProphet, which uses protein grouping to also adjust the peptide probability.

11.5.2 Scaffold

Scaffold (http://www.proteomesoftware.com/) is a comprehensive package for processing of proteomic and metabolomics data. This software is compatible with all major search outputs such as Mascot, Sequest, and Andromeda, and accommodating packages such as Proteome Discoverer Spectrum Mill, OMSSA, Tandem and, less used, IdentityE, MSAmanda. Scaffold provides tools for analysis of quantitative proteomic experiments such as iTRAQ, TMT and SILAC. Data visualization as well as visualization (Q+ component) and quantitative (Q + S component) site location of posttranslationally modified proteins is possible.

11.5.3 ProteoIQ

ProteoIQ (http://www.premierbiosoft.com/protein_quantification_software/index.html), like Scaffold, is designed for statistical validation, protein quantification and comparative proteomics. It supports integration of data outputs from other search tools such as Mascot, SEQUEST, or X! Tandem. ProteoIQ incorporates a false discovery rate and probability score. It can also perform quantitative analysis for spectral counting, precursor intensity, iTRAQ, TMT, and SILAC workflows. Users can link to an external database such as Swissprot or an internal database such as LIMS.

11.5.4 Skyline

Unlike the previously mentioned software, Skyline (https://brendanx-uw1.gs.washington.edu/labkey/project/home/software/Skyline/begin.view) is open-source software for taking spectral data and sequence information and building targeted proteomics methods, such as MRM and DIA/SWATH. There is a growing trend toward using MRM to validate proteomics experiments and thus the number of targeted analyses needed. Skyline helps streamline the process for both large and small studies.

11.6 Searches for Posttranslational Modifications

Posttranslational modifications (PTM) occur in different biological contexts and complex protein mixtures extracted from biological samples invariably contain those carrying PTMs. Another level of complication is that the overall PTM can be heterogeneous, consisting of various modifications present on the same peptide. One such example is histones, which have a mosaic of PTMs dynamically changing in time as a result of their regulatory role [19]. For an effective search of peptides with PTMs, an exact mass of such modification is added to search parameters along with appropriate mass accuracy parameters.

PTMs might be fixed or variable. A fixed modification is expected to be present in every peptide having a specific amino acid. The best example is alkylation of cysteine residues, which occurs during reduction and alkylation of protein prior to enzymatic digestion. Variable modifications may or may not be present, such as phosphorylation of serine, threonine and tyrosine occurring in a percentage of the population of protein molecules. If variable phosphorylation modification on these residues is specified, the search engine will test for a match with the experimental data where the phosphorylation events may or may not occur. Searching for variable modifications is a powerful tool for finding out the PTMs; however, one needs to be cautious when specifying the number of variable modifications since adding even a single variable modification will generate more possible peptides to be searched against. If there are multiple modifiable residues within a single peptide, the workload for searching all the possible modification permutations could be exponentially increased. This could drastically increase the search time and decrease the discriminating power of the search. Not every algorithm is suitable for searching large datasets for PTMs; however, we expect rapid development in this area because of the importance of PTMs in biological processes requiring more precise detection and quantification.

11.7 Summary

Along with development of general databases, highly focused resources are being created, such as databases compiling current status of protein phosphorylation, etc. It is impossible to comprehensively review all these resources within the limitations of this chapter; therefore we focused here on the most basic and most widely used tools and resources, urging readers to seek information on new resources and releases of new versions. This is a very dynamic process as new information becomes available daily. Because of the increasing size of proteomic datasets and multiple tools that are needed, we expect that in the near future all such analyses will likely be performed in the computing cloud.

Protein databases are constantly changing with the continuous process of annotation, integration of information originating from various types of experiments such as crystallography, posttranslational modifications, biologically relevant mutations, etc. Organization of information is becoming more and more user-friendly despite expansion. Nevertheless, information included in protein databases is not complete and it is impossible to estimate when, if ever, we will be able to claim victory of completeness. Despite this, it is highly advisable to verify information obtained from proteomic experiments before anybody claims its uniqueness. Once more, we need to emphasize the absolute necessity of referencing versions, releases and dates of protein resources while reporting results. If everybody adheres to such rules, we collectively will spend less time comparing findings of complex proteomic experiments.

References

[1] Consortium U. The Universal Protein Resource (UniProt). Nucleic Acids Res 2007;35:D193−7.
[2] Wu CH, Apweiler R, Bairoch A, Natale DA, Barker WC, Boeckmann B, et al. The Universal Protein Resource (UniProt): an expanding universe of protein information. Nucleic Acids Res 2006;34:D187−91.
[3] Suzek BE, Huang H, McGarvey P, Mazumder R, Wu CH. UniRef: comprehensive and non-redundant UniProt reference clusters. Bioinformatics 2007;23:1282−8.
[4] Maruyama Y, Kawamura Y, Nishikawa T, Isogai T, Nomura N, Goshima N. HGPD: Human Gene and Protein Database, 2012 update. Nucleic Acids Res 2012;40:D924−9.
[5] Rose CM, Venkateshwaran M, Grimsrud PA, Westphall MS, Sussman MR, Coon JJ, et al. Medicago PhosphoProtein Database: a repository for Medicago truncatula phosphoprotein data. Front Plant Sci 2012;3:122.
[6] Yates 3rd JR, Eng JK, McCormack AL, Schieltz D. Method to correlate tandem mass spectra of modified peptides to amino acid sequences in the protein database. Anal Chem 1995;67:1426−36.
[7] Mann M, Wilm M. Error-tolerant identification of peptides in sequence databases by peptide sequence tags. Anal Chem 1994;66:4390−9.
[8] Pappin DJ, Hojrup P, Bleasby AJ. Rapid identification of proteins by peptide-mass fingerprinting. Curr Biol 1993;3:327−32.

[9] Shilov IV, Seymour SL, Patel AA, Loboda A, Tang WH, Keating SP, et al. The Paragon Algorithm, a next generation search engine that uses sequence temperature values and feature probabilities to identify peptides from tandem mass spectra. Mol Cell Proteomics 2007;6:1638–55.

[10] Cox J, Mann M. MaxQuant enables high peptide identification rates, individualized p.p.b.-range mass accuracies and proteome-wide protein quantification. Nat Biotechnol 2008;26:1367–72.

[11] Cox J, Neuhauser N, Michalski A, Scheltema RA, Olsen JV, Mann M. Andromeda: a peptide search engine integrated into the MaxQuant environment. J Proteome Res 2011;10:1794–805.

[12] Craig R, Beavis RC. TANDEM: matching proteins with tandem mass spectra. Bioinformatics 2004;20:1466–7.

[13] Aebersold R, Mann M. Mass spectrometry-based proteomics. Nature 2003;422:198–207.

[14] Hardman M, Makarov AA. Interfacing the orbitrap mass analyzer to an electrospray ion source. Anal Chem 2003;75:1699–705.

[15] Hu Q, Noll RJ, Li H, Makarov A, Hardman M, Graham Cooks R. The Orbitrap: a new mass spectrometer. J Mass Spectrom 2005;40:430–43.

[16] Schwartz JC, Senko MW, Syka JE. A two-dimensional quadrupole ion trap mass spectrometer. J Am Soc Mass Spectrom 2002;13:659–69.

[17] Clauser KR, Baker P, Burlingame AL. Role of accurate mass measurement (+/− 10 ppm) in protein identification strategies employing MS or MS/MS and database searching. Anal Chem 1999;71:2871–82.

[18] MacLean B, Tomazela DM, Shulman N, Chambers M, Finney GL, Frewen B, et al. Skyline: an open source document editor for creating and analyzing targeted proteomics experiments. Bioinformatics 2010;26:966–8.

[19] Burlingame AL, Zhang X, Chalkley RJ. Mass spectrometric analysis of histone posttranslational modifications. Methods 2005;36:383–94.

12

DESIGN AND STATISTICAL ANALYSIS OF MASS-SPECTROMETRY-BASED QUANTITATIVE PROTEOMICS DATA

F. Yu, F. Qiu and J. Meza

University of Nebraska Medical Center, Omaha, NE, United States

CHAPTER OUTLINE

12.1 Introduction   212
12.2 Mass Spectrometry-Based Quantitative Proteomics   213
 12.2.1 Stable Isotope Labeling   213
 12.2.2 Label-Free Quantification   214
12.3 Issues and Statistical Consideration on Experimental Design   214
 12.3.1 Randomization   215
 12.3.2 Technical Replicate or Biological Replicate   215
 12.3.3 Experimental Layout and Label Assignment   216
 12.3.4 Label-Free Experimental Layout   217
 12.3.5 Stable Isotope Labeling Experiment Layout   219
 12.3.6 Sample Size Calculation   221
12.4 Data Preprocessing for Statistical Analysis   224
 12.4.1 Data Preparation and Filtering   224
 12.4.2 Transformation   225
 12.4.3 Normalization   225
 12.4.4 Missing Value Imputation   226
12.5 Statistical Analysis of Protein Expression Data   227
 12.5.1 Differentially Expressed Proteins   228
 12.5.2 Time-Dependent Proteins   230

Proteomic Profiling and Analytical Chemistry. http://dx.doi.org/10.1016/B978-0-444-63688-1.00012-4

12.5.3 Multiple Comparisons Across Proteins, and False
Discovery Rate 230
12.5.4 Clustering 232
12.5.5 Principal Component Analysis 233
12.5.6 Protein Networks 233
12.6 Summary 234
References 235

12.1 Introduction

Proteomics is the large-scale study of proteins, particularly their structures and functions. A main goal of these proteomics studies has been the quantitative analysis of the proteome of a species or a particular cell or tissue type. Recent advances in molecular and computational biology have allowed for development of powerful high-throughput techniques to examine protein expression at the cellular level. One widely used high-throughput technique is mass spectrometry (MS). Although high-throughput techniques for proteomics studies provide rich information on biological processes, they can be costly in terms of equipment, consumables and time. Therefore, careful experimental design is critically important for proteomics studies to make full use of the available resources and efficiently answer the questions of interest. In addition, proteomics studies generate hundreds to tens of thousands of protein/peptide sequences. The determination of the abundance of a large number of protein/peptides followed by analysis of the protein expression presents great computational and statistical challenges for data analyses. An appropriate data analysis method should fit the characteristics of the proteomics studies and the experimental design, as well as provide an accurate answer to the question of interest. In this chapter, we will first describe a couple of widely used MS-based quantitative proteomics experiment types. Following that, we will discuss the concepts and challenges for experimental design and statistical analysis of proteomics data for each type of quantitative MS-based proteomics study.

12.2 Mass Spectrometry-Based Quantitative Proteomics

MS has been widely used for quantitative proteomics to quantify the absolute or relative protein expression levels from different biological conditions. The workflow of the MS-based proteomics experiments can be classified into two categories: stable isotope labeling and label-free quantification. Analysis of data-independent acquisition data such as SWATH-MS is described in detail in a separate chapter.

12.2.1 Stable Isotope Labeling

Stable isotope labeling has been commonly employed in many spectrometry-based quantitative proteomics experiments [1]. The biological samples are labeled by different isotopes, mixed together, and digested into peptides. As different isotopes have different masses, the samples from which the peptides were extracted are recognized by the mass spectrometer and the abundance of the peptides from each sample is quantified. The isotope labeling can be attached to the amino acid metabolically or chemically. The metabolic-labeling method incorporates the isotopic labels during the process of cellular metabolism and protein synthesis. One popular metabolic-labeling method is stable isotope labeling with amino acids in cell culture (SILAC) [2], which metabolically incorporates "light," "medium" or "heavy" forms of amino acid into the proteins, and allows simultaneous quantification of proteins from three cellular states. Chemical labeling is an important alternative isotope-labeling technique with its own advantages. The isotope-coded affinity tag (ICAT) method [3] and the isobaric tags for relative and absolute quantification (iTRAQ) [4] are two widely used chemical isotope-labeling methods. The ICAT reagents consist of three parts: a cysteine-reactive group, an isotopic light or heavy linker and a biotin affinity tag. The cysteine residuals of proteins from two different biological samples can be labeled by the light or heavy ICAT reagents. Quantification of proteins from these two ICAT reagent-labeled samples can be obtained. In the iTRAQ experiment, a reagent consists of an amine-reactive group, a balance group and a reporter group. The reporter groups have eight different masses ranging from 114 to 121. Each reporter group is mass matched with its own balance group and creates different isoforms with identical masses. The iTRAQ experiment allows multiplex quantification for up to eight samples, and is particularly useful for studying time-dependent proteins or protein expression under multiple biological conditions.

12.2.2 Label-Free Quantification

Label-free quantification applies two quantification strategies: (1) spectral counting or (2) spectrometric signal intensity to measure the protein expression. Each sample will be analyzed by the mass spectrometer separately using the same protocol. The proteins from each sample are identified, and the protein expression from each sample is estimated using either the number of MS/MS spectra identifying peptide of the protein or the intensity of the corresponding MS spectrum features of the protein. Although proteomics experiments may be conducted using different MS-based proteomics profiling techniques, their output shares a similar structure including the list of detected proteins and the absolute or relative abundance of the proteins across all samples for each experimental run.

12.3 Issues and Statistical Consideration on Experimental Design

Experimental design involves a complex procedure including defining the population of interest, selecting the individuals (samples) from the underlying population, identifying the number of samples needed, allocating those samples to different

biological conditions, and planning the data acquisition based on available resources. Careful experimental design must occur before any data are
collected to assure success in data acquisitions and
avoid unnecessary waste of time and resources. In
the following sections, we will review the important
issues and concepts for experimental design.

12.3.1 Randomization

Randomization is important for experimental
design of proteomics experiments. First, the
samples should be randomly selected from the
population, so that the inference using the sample
data can be generalized to the population. More
importantly, the use of randomization can avoid
bias caused by potentially unknown systematic
errors. For example, when the data acquisition
cannot be completed at the same time, the sample
processing and data acquisition should be randomly
run at each time so that the potential effects from
the experimental conditions will equally influence
the experimental data. In this way, potential confounding of time with biological condition can be
avoided.

12.3.2 Technical Replicate or Biological Replicate

Proteomics experiments require multiple replicates of measurements to ensure the reproducibility
of the results. Replication is classified into two types:
technical replicates and biological replicates. Technical replicates are repeated measures from the same
biological samples that allow measurement of the
error of the experimental techniques. Technical
replicates also increase the quality of the measurements on the same sample. Biological replicates are
multiple measurements from different biological
samples of the same biological condition. Although
randomization will ensure measurements from
different treatment conditions to be as similar as
possible, measurements from different samples of
the same treatment conditions will contain variation

due to difference in individual sample characteristics (eg, environmental factors). Biological replicates from randomly selected samples of the same biological conditions will help the user to assess whether the observed differences in the measurements exist due to the involvement of different biological conditions instead of random chance. Biological replicates generally are more important than the technical replicates, as proteomics experiments usually focus on identifying the difference associated with the treatment rather than differences between samples of the same biological conditions or technical errors.

12.3.3 Experimental Layout and Label Assignment

Experimental layout is another important component of experimental design. The proteomics experiment is usually conducted to quantify and compare the protein expression level of multiple biological conditions. In the labeling of proteomics samples, multiple labels will be used to identify and measure the protein expression levels simultaneously. For example, the pSILAC experiment can label three different biological samples while the eight-plex iTRAQ experiment can label eight different biological samples. The generation of multiple replicates will require multiple experimental runs. It is important to avoid the confounding effects from both label and experimental runs. The experimental layout will help to avoid bias by addressing which experimental conditions were hybridized on the same experimental run and which will be labeled for certain biological conditions. A key in the experimental layout is to assign each biological condition to each label and each experimental run with similar probability. In addition, variation between measurements from different experiments is generally larger than measurements from the same experiment. By optimizing the method of sampling from different biological conditions between different experimental runs, the experimental layout can improve the efficiency of the design and analysis.

Depending on the experiment type and the number of experimental conditions relative to the number of labels, we can consider the following options for experiment layout and label assignment.

12.3.4 Label-Free Experimental Layout

In the label-free experiment, the main focus of the layout is to avoid bias from different experimental runs. An equal number of samples per biological condition can be sampled under each experimental run. The samples from each biological condition should be randomly sampled to control bias that may arise from potential environmental conditions (ie, the data acquisition time) as mentioned in the section on randomization.

Randomized complete block designs (RCBD) are recommended for label-free proteomics experiments. A block usually is defined based on an important factor in the effect of the biological conditions. Specifically, samples evaluated by one experimental run are considered as a block. Then, samples from the experimental run are randomly chosen from each biological condition once and only once. Note that the RCBD allows one sample from each biological condition to be evaluated by the same experimental run. In this way, the confounding effects from the experimental run will be washed out in the comparison between biological conditions. One example of RCBD layout for four biological condition groups is shown in Fig. 12.1A to compare the protein expression levels from all four biological condition groups. Specifically, four samples each from the four treatment options can be evaluated using four experimental runs via the RCBD. The difference in the protein expression levels from the four treatment options can be estimated using the average expression level from the treatment options across the four experimental runs. Since the comparison within the same experimental run will not be affected by the variation between experiments, the estimates for the difference in the protein expression levels from any pair of the four treatment options will not be affected by the variation between different experimental runs.

(A) Randomized Complete Block Design

E1	D1	D2	D3	D4
E2	D1	D3	D2	D4
E3	D2	D3	D4	D1
E4	D3	D4	D1	D3

(B) Latin Square Design

	T1	T2	T3
E1	D1	D2	D3
E2	D2	D3	D1
E3	D3	D1	D2

(C) Balanced Incomplete Block Design

	T1	T2	T3
E1	D1	D2	D3
E2	D1	D2	D4
E3	D1	D3	D4
E4	D2	D3	D4

(D) Three Label Reference Design

	T1	T2	T3
E1	C(ref)	D1	D2
E2	D2	C(ref)	D1
E3	D1	D2	C(ref)
E4	C(ref)	D3	D4
E5	D4	C(ref)	D3
E6	D3	D4	C(ref)

(E) Three Label Loop Design

	T1	T2	T3
E1	D1	D2	D3
E2	D4	D3	D2
E3	D3	D4	D1
E4	D2	D1	D4

Figure 12.1 Five experiment layouts. (A) is plotted for label-free experiments studying four biological conditions; hence no label assignment is involved. (B)–(E) are plotted for three-label workflow. (B) uses a Latin square design, hence only three biological conditions are able to be studied. (C)–(E) each study four biological conditions using balanced incomplete block design, reference design and loop design, respectively. (E1 (E4) denotes the first (fourth) experimental run; D1 (D4) denotes the first (fourth) biological group; C denotes the control group considered as a common reference; T1 (T3) denotes the first (third) label.)

12.3.5 Stable Isotope Labeling Experiment Layout

Different from the label-free experiment, an experiment labeling the proteins using stable isotopes may introduce bias to the quantification of the protein expression levels due to different efficiency in labeling and hybridization. The simplest design to reduce label bias uses label swapping.

12.3.5.1 Label Swapping

Consider a simple experiment setting with only two treatment conditions, in which the experimental design uses two different labels for protein quantification. A concern with this approach is that observed differences between the two biological conditions may be due to the different labeling used for the samples. The label swapping design can be used to reduce this label bias. In this design, two experimental runs will be used for replicates. Under one experimental run, two labels will be randomly assigned to the biological conditions; and under another experiment the labels will be switched to label these two biological conditions. Since the samples from different biological conditions will be randomly assigned to the label for sampling under each experimental run, the confounding effects from labeling and different experimental runs will be washed out when evaluating the relative expression levels of the proteins between different biological conditions.

12.3.5.2 Latin Square Design

The Latin square design is a general version of the dye-swapping design for samples from more than two biological conditions. The Latin square design requires that the number of experimental conditions equals the number of different labels. The same number of experimental runs as the number of treatment conditions is also used. The treatment conditions are labeled once using each label and sampled once under each experimental run. Fig. 12.1B shows one way of experiment layout when Latin square design is used for three-label experiments studying the protein expressions under three biological conditions. The advantage of the Latin square design is to control the variation from different labels and different experimental runs. The Latin square also provides better efficiency than the RCBD [5].

12.3.5.3 Balanced Incomplete Block Design

In some complex scenarios, the number of biological conditions to be compared exceeds the number of labels available per experimental run. In the balanced block design, each experimental run can be considered as block. When all biological conditions cannot be assigned to the same block (or experimental run), the balanced incomplete block design can be used. In this design, a minimum number of blocks (or experimental runs) are be used so that all pairs of biological conditions appear together in the same block (or experimental run) an equal number of times (Fig. 12.1C). Under each experimental run, the labels are randomly assigned to the biological conditions.

When the block size is small (that is, the number of labels per experimental run is small), instead of using a balanced incomplete block design, other specialized design options previously proposed for microarray experiments (Kerr et al. [6], Dobin and Simon [7], Woo et al. [8]) can be used for designing the experimental layout for proteomics experiments. In the following sections, we illustrate two popular design options.

12.3.5.4 Reference Design

The reference design uses a common reference sample under each experimental run to control the between-experiment variation. The reference sample usually is not of interest, yet it facilitates the comparison between the samples from different experimental runs. Usually we can select samples from the normal control group as the reference when evaluating the protein expression levels associated with treatment. When possible, it is recommended to utilize label swapping between samples of interest within the same block. Kerr and Churchill [9] stated that reference designs coupled with direct dye-swaps between samples of interest can result in powerful, robust and readily extendible sets of comparisons. Fig. 12.1D provides an example of reference design for an experiment studying four biological conditions. The condition C is sampled as a common reference and was sampled under all experimental runs.

12.3.5.5 Loop/Cyclic Design

The loop/cyclic design is a special type of the balanced incomplete-block design in which each pair of biological conditions will be assigned to the same block with the same frequency. To develop the loop/ cyclic design, the biological conditions are randomly assigned to different labels in the first experimental run. Then the labeling order of the experimental conditions in the first experimental run are cycled and used for label assignment in the subsequent experimental runs. The cyclic permutation is continued until all biological conditions have been sampled and the desired number of comparisons has been made. Fig. 12.1E provides an example of loop design using three labels studying four biological condition groups.

12.3.6 Sample Size Calculation

Sample size calculation is an important aspect of experimental design. The goal is to calculate the number of replicates for the experiments being planned. The number of replicates should be large enough to ensure that the proteomics experiment will have adequate power to address the question of interest while not being so large that it is inefficient in terms of time and cost. In this chapter, we focus on sample size calculations for identifying differentially expressed proteins using high-throughput proteomics data.

Due to the importance of sample size calculation in experimental design, there is an abundance of literature published that addresses sample size calculations for different types of experiments. Campbell et al. [10] summarize the sample size calculation approaches for studies involving continuous, binary or ordered categorical outcomes for clinical trials. Tibshirani [11] proposed sample size calculations for identifying differentially expressed genes using microarray data using a permutation t test. Dobbin and Simon [7] focused on disease status prediction using microarray data and proposed sample size calculations for microarray data. Cairns et al. [12] proposed sample size calculations for identifying differentially expressed proteins using proteomics data. As with more conventional experimental designs, proteomics experiments seek to compare continuous data between groups in order to detect differentially expressed proteins. However, the complexity of the proteomics experiment and the availability of a large amount data complicate the sample size calculations. For example, the proteomics experiment might identify multiple peptide sequences matched for the same protein via multiple spectrum runs. Therefore, multiple observations may be available for the same protein per biological sample under the same experiment. In addition, the comparison of protein expression levels will occur on each protein, which raises issues with multiple comparisons thus inflating the Type I error rate. Thus, a conservative threshold for the Type I error rate is required for proteomics studies as compared to statistical tests involving only hypothesis.

First, we investigate how multiple comparison issues affect the sample size calculation results. As mentioned previously, the multiple comparison issues requires a conservative threshold for Type I error rate. Instead of directly controlling the Type I error, the analyst usually estimates the false discovery rate (see Section 12.6.3) and identifies differentially expressed proteins controlling for the false discovery rate. To calculate the sample size for proteomics studies, it is helpful to identify the value for the Type I error rate given a prespecified false discovery rate. Assume that the data from all proteins share the same distribution and the same statistical test will be conducted on all proteins. Also assume that G proteins are composed of m_0 equally expressed proteins and m_1 differentially expressed proteins. If the statistical test controls for the average Type II error rate to be less than β, and the false discovery rate to be less than q, Banjamini and

Hochberg [13] inferred that the corresponding average Type I error will be controlled at:

$$\alpha_{\text{avg}} \leq (1 - \beta)_{\text{avg}} \frac{q}{1 + (1 - q)m_0/m_1}.$$

For example, assume the statistical test is set to control the false discovery rate to be less than 0.05, and the average power $(1 - \beta)$ to be 0.8. Assume that 5% of proteins are truly differentially expressed, which ensures the ratio $m_0/m_1 = 95/5 = 19$. Then using the equation above, the controlled average type I error rate is 0.0021, which is much smaller than 0.05. Then, the sample size calculated for the proteomics study will approximately equal the number of replicates required by one protein with a type I error rate of 0.0021 and a target power of 0.8, which can be calculated following the sample size calculation approaches for single hypothesis testing using standard statistical software.

Note that the sample size calculation is conducted to optimize the power of detecting differential expression between biological conditions. We assume that there are two biological conditions in comparison, and the log-transformed data measuring the protein expression levels are normally distributed. The variance of the evaluated differential expression levels need to be estimated for sample size calculation. When a proteomics experiment has multiple measurements from the same biological sample on the same protein, the repeated measurements will be considered as technical replicates, and the estimated variance is expected to be smaller with more technical replicates. The estimate for the variance also depends on the planned experimental layout. More detailed discussion on the experimental layout and the corresponding variance estimates was given in Oberg and Vitek [14,15]. For a SILAC experiment with label swapping design, a one-sample t-test can be used to evaluate the differential protein expression between the treated group and the untreated control group. Note that the sample size calculation can be conducted via formula:

$$n = \frac{\left(Z_{\alpha/2} + Z_\beta\right)^2 \sigma^2}{d^2},$$

where Z_β is the z-score from the standard normal distribution such that the area to the right of Z_β is β, d is the difference in means we wish to be able to detect, and σ is the outcome standard deviation. Assume that 5% of proteins are truly differentially expressed. We need to control the Type I error to be less than 0.0021 to ensure a control of the FDR at 0.05, and an average power of 0.8. The formula for the sample size is:

$$n = \left(Z_{0.0021/2} + Z_{0.8}\right)^2 0.6^2 \Big/ 1^2 = 6.$$

Thus, we need six experimental runs to identify a true twofold change with an average power of 0.8 when controlling the false discovery rate at 0.05.

12.4 Data Preprocessing for Statistical Analysis

Raw data from quantitative proteomics experiments usually are not ready for statistical analysis. Similar to the gene expression data analysis, a series of data preprocessing procedures are first applied to prepare the data for the further statistical analysis.

12.4.1 Data Preparation and Filtering

Prior to any statistical analysis, the MS spectra data will be processed by peptide searching software. The protein/peptide sequence will be identified and the abundance measures of the protein will be quantified. The software usually assesses the confidence of the identified peptides and proteins. For example, the Protein Pilot (v5.0) [16] software can be applied to processing iTRAQ experimental data, and a measure named "confidence" to quantify the probability that the hit in the peptide searching is a false positive can be calculated. SILAC experimental data can be processed using MaxQuant [17] to identify peptides and assemble peptides into proteins. MAXQuant also calculates the false discovery rate for both protein and peptide identification. The statistical analysis should then be focused on the data from proteins and peptides with adequate identification accuracy. In addition, protein assembled from only one peptide match should be excluded from further analysis. A more detailed review of the open source libraries and frameworks for data processing and quantification of MS-based proteomics experiments can be found in Perez-Riverol et al. [18].

12.4.2 Transformation

Raw data from quantitative proteomics experiments generally are not normally distributed, which prevents the use of many commonly used statistical methods due to the violation of the normality assumption. Therefore, protein expression will be transformed so that the transformed values satisfy the normally assumption. The log-transformation is the most commonly used transformation and is also helpful to stabilize the variances of the protein expression values, particularly for the experiments with larger variances for proteins of high expression values, and smaller variances for proteins of low expression values.

12.4.3 Normalization

Proteomics experiments usually come in replicates in order to reduce the variation from the biological system or experimental conditions. Normalization is an important data preprocessing step for replicated proteomics experiments. Note that it is hard to avoid the technical effects from the proteomics experiments due to sample mixing errors, incomplete isotope incorporation, or isotope impurity. The existence of technical effects may cause underestimation on the underlying effects of biological conditions. A carefully designed experiment will improve the data quality by reducing the confounding effects of the experimental settings (for example, labeling). The normalization procedure provides an additional important approach to reduce the estimation bias due to the technical effects. Normalization usually begins with a calibration procedure, which will ensure that data from different experiments of the same biological conditions share a similar center value. Then, differences between biological conditions are attributed to the effects of the biological condition, instead of the technical bias. For example, data from the metabolic labeling pSILAC proteomics experiments will be collected in the form of ratios between samples from two biological conditions labeled by different medium. The Max-Quant (version 1.5.1.2) will calibrate the medium log2 ratio values from the pSILAC data from different experiments to zero, as it assumes that most proteins are not differentially expressed. The calibration method also efficiently reduces the label effects when samples from different biological conditions are labeled differently. For some studies with more complex experimental settings, the normalization method can be more sophisticated. For the iTRAQ replicate experiment, protein expression may incorporate variations from animal, protein, peptide and experimental condition. Accordingly, Oberg et al. [19] proposed an iterative back-fitting procedure on log-transformed protein expression levels to remove the animal, protein and peptide effects. The SAS/STAT® software for Windows version 9.2 or higher code "PROC itraqnorm" was made available online [20] by Douglas Mahoney at the Mayo Clinic for the back-fitting procedure.

12.4.4 Missing Value Imputation

Missing value imputation is another important preprocessing step. In proteomics experiments, peptides will be randomly sequenced by mass spectrometer and only a subset of proteins present can be identified. When multiple proteomics experiments are used for protein quantification, many identified proteins fail to be quantified in all experiments. The incompleteness of protein identification and quantification introduces a great number of missing data to the proteomics raw data, and requires a scientific method for handling missing data. However, most statistical methods assume complete data. A simple analytical approach is to focus only on the proteins with complete protein quantification information so that standard statistical methods may be used. However, this will result in a loss of information due to excluding incomplete data. An alternative method is to impute the missing values based on the available values using average expression value of the same protein. More sophisticated methods use available values from other related proteins to impute the missing value [21]. To handle missing data for iTRAQ experiments, Luo et al. [22] assumed the data are not missing at random and proteins with lower abundance values are more likely to be missing. A hierarchical Bayesian approach was proposed to fit both missing data and observed data to evaluate the protein expression levels from multiple iTRAQ experiments [22]. For the label-free liquid chromatography-mass spectrometry (LC-MS) proteomics data, Webb-Robertson et al. [23] provided an evaluation of several commonly used imputation strategies and found that local similarity-based imputation methods had better performance than naive and global imputation methods in terms of classification accuracy.

12.5 Statistical Analysis of Protein Expression Data

High-throughput proteomics experiments measure the expression of thousands of proteins simultaneously using samples from a number of biological conditions. Such experiments produce high-dimensional data sets, which may be influenced by large variation from biological, technical and experimental factors. To utilize an appropriate data analysis method, researchers need to obtain a comprehensive understanding of the experimental design prior to conducting any data analysis. The most important step is to determine the aim of the experiment. Proteomics experiments usually are designed for two types of studies: (1) the interrelation between the proteomics expression and certain sample groups (for example, is there differential expression of a protein between different treatment groups? Does a protein exhibit time-depending change?); and (2) the dependencies between proteins (for example, do the proteins share similar patterns?). In this chapter, we will focus on the statistical methods to address these two types of studies.

12.5.1 Differentially Expressed Proteins

One of the most common aims of proteomics experiments is to explore the interrelation between the proteomics expression and certain sample groups. A simple example is the comparison of protein expression profiles in two or more different types of biological conditions such as untreated control and treated group. One experiment typically collects a certain number of expression levels from each biological condition. On each protein, a pair of hypotheses is set up with null hypothesis: (1) the protein is equally expressed or there is no difference in the protein expression values from different biological conditions, or (2) an alternative hypothesis: the protein is differentially expressed or there is a difference in the protein expression between some biological conditions. The differential expression in proteins between any two biological conditions will be estimated using the fold change, which equals the ratio between the geometric mean values of the expression levels from the two biological conditions in comparison. The hypothesis testing will be conducted on each protein to evaluate whether there is enough evidence to reject the null hypothesis, hence concluding that the protein is differentially expressed. The statistical tests will be highly dependent on the experimental settings. For example, when the data are normally distributed, the two-sample t-test can be applied for hypotheses testing when there are two groups to be compared on the same protein, while the analysis of variance method (ANOVA) can be used when more than two groups are being compared. When the data are not normally distributed, nonparametric methods such as a Mann–Whitney test or Kurskal–Wallis test can be used. In addition, a multiple-factor ANOVA or linear model can be used to adjust for the confounding effects of potential experimental factors (ie, different disease stages) that affect the protein expressions. An application was described in Wiederin et al. [24] where the iTRAQ experiment was used to study protein expression levels from three biological conditions: baseline, acute infection and chronic infection. The iTRAQ data was first log-transformed and normalized using the iterative back-fitting procedure to remove the confounding effects of the experimental conditions. The normalized data were compared using nonparametric methods since the normalized data were not normally distributed. In another application for SILAC data [25], the uninfected control cells were labeled with "medium" media and the HIV-infected cells were labeled with "heavy" media. The fold changes measuring the relative protein expression levels between the heavy media labeled samples and those labeled with medium media were calculated. The significant B values [17] were further calculated with the test statistic equal to the ratio between the log2 fold change and the estimated standard deviation of the log2 fold change. The proteins were identified to be differentially expressed between the HIV-infected samples and the uninfected samples if they have small B values.

In proteomics experiments with multiple conditions, we may be interested in comparisons among only a certain subset of conditions. Depending on the questions to be addressed, we can use two different approaches to construct the hypotheses testing. For example, if we are interested in a set of related comparisons to examine the overall evidence of differential expression for the studied proteins, then a global test (ie, F-test in the ANOVA method) can be used to evaluate whether there is differential expression from any considered comparisons. In another scenario, when the comparisons in consideration are formed to answer independent questions, then separate analyses on each protein can be used to answer each comparison independently. When multiple comparisons are conducted on one protein, additional multiple comparison procedures may be needed to adjust for the inflation of type I error from multiple comparisons.

Considering that the high-throughput proteomics experiment will collect expression levels from multiple proteins in parallel, Bayesian methods have been proposed to construct parallel models on each protein. The parameters of the parallel models share a common prior distribution to borrow information across different proteins. Bayesian methods can be applied to experiments with two conditions or multiple conditions [22,26–28]. Based on different statistical methods, the p value or the posterior probability for the protein having unequal expression levels between biological conditions will be used to evaluate whether a significant difference exists in the protein expression levels between biological conditions.

12.5.2 Time-Dependent Proteins

Proteomics experiments have been used to study the change in the protein expression levels across time. For example, the protein expression levels were measured from samples at different time periods. Different from proteomics experiments conducted at a fixed time point, proteomics experiments studying the time trend will focus on evaluating the time effects, the treatment effects and the interaction between the time and treatment. A protein with significant treatment–time interaction implies that association between treatment and protein expression differs by time. Analysis of variance is often used for studying time-dependent protein expression values. However, when samples are collected from the same patient at different time-points, a repeated measures analysis of variance should be used to properly account for the correlation among measurements collected from the same patient.

12.5.3 Multiple Comparisons Across Proteins, and False Discovery Rate

Note that the process of detecting differentially expressed proteins involves hypotheses testing on multiple proteins. For each comparison on a single protein, two types of error may occur. A Type I error occurs when a protein without differential expression is incorrectly declared to be differentially expressed. A Type II error occurs when a protein is declared to not be differentially expressed when in fact there is differential expression. Since each hypothesis test can result in an error, multiple hypotheses testing for a large number of proteins will drastically inflate the overall Type I error. Standard statistical methods control the overall Type I error to be less than 0.05. However, proteomics data require a more conservative Type I error threshold.

Two error rates are defined to evaluate the overall error rate for comparison over all proteins. The first is the family-wise error rate (FWER), which measures the probability of at least one false positive among all comparison. For example, Westfall and Young [29] proposed a step-down maxT permutation adjustment on p-values from genes or proteins to control the FWER. The second is the false discovery rate (FDR), which measures the false positive rate among the rejected hypotheses (the detected differentially expressed proteins). We note that the FDR is less stringent than the FWER [13], and hence is commonly used for addressing multiple comparison issues in proteome studies. For example, Storey and Tibshirani [30] proposed a permutation t-test to estimate the FDR. Efron [31−34] provides a gene-specific measure called local FDR to bound the global FDR and to estimate the false negative rate. Storey [35] defined q value for each gene to measure the proportion of false positive that occurred (or FDR) when the gene is called differentially expressed. Pounds and Morris [36] fit a beta-uniform mixture distribution on the p values across all genes or proteins to estimate the FDR.

In this section, we will use the Benjamini Hochberg (BH) method [13] to illustrate how this method can be used to control the false discovery rate and identify differentially expressed proteins. First, an appropriate statistical analysis has been conducted for each protein to evaluate the differential protein expression among the biological conditions of interest. From this analysis, we obtain a p-value for each of the G proteins that are differentially expressed, denoted as p_g for the gth protein with $g = 1, \ldots, G$. Then, the p values are ordered as $p_{r1}, p_{r2}, \ldots p_{rG}$, in ascending order, where r_g is the protein ID ranked in the gth position in the sorted list based on the calculated p values. Let k be the largest integer i for which $p_{ri} < (i/G) \times \alpha$ for all i. Then we declare all the proteins with labels $r_1; \ldots ; r_k$ to be differentially expressed. Note the value minimum $p_{r_g} \times (G/g)$ for $g = i, i+1, \ldots, G$ for the protein with the ith smallest p value is referred as the adjusted p value. The BH procedure [13] uses a sequential p value method so that on the average FDR $< \alpha$ for some prespecified α.

12.5.4 Clustering

Another important experimental aim of proteomics experiments is the dependency between proteins. Cluster analysis assumes that proteins with similar biological functions share similar protein expression patterns, and subdivides the proteins in different clusters, so that proteins in the same cluster share similar patterns of protein expression levels compared to proteins in different clusters.

Hierarchical clustering [37] is one clustering analytical method to build clustering in a hierarchy, providing additional insight on the dependencies between the proteins under study. First, distance metrics are used to evaluate the similarity between proteins based on their protein expression levels. A common choice for the distance metric is the Euclidean distance or the correlation coefficient. The expression data may need to be standardized to have a mean of 0 and a standard deviation of 1 for these metrics. The distance between two protein sets can be measured by summarizing the distance measures between any paired proteins from these two protein sets using a link function. For example, the average link function will use the average distance between two proteins from these two protein sets to evaluate their similarity. First, the distance between any two proteins is calculated, and the protein pair with minimum distance is connected using the same "branch." The procedure will be repeated between the protein sets containing the connected proteins until all proteins are connected by a branch. As a result, in the formed hierarchy structure, the protein sharing similarity in the expression levels will be linked by the same branch, with the length of branches implying the strength of similarity.

12.5.5 Principal Component Analysis

Principal component analysis (PCA) [38] is a widely used statistical procedure on mass-spectrometry data for dimension reduction and clustering visualization. Specifically, the principal component analysis will use an orthogonal transformation to identify principal components, which equal a linear combination of the protein levels and are linearly uncorrelated with each other. The identified principal components are expected to account for most of the variability of the data from different samples. Therefore, the PCA can use the identified principal component to distinguish sample or protein subsets that are responsible for the majority of the variations between groups, and effectively reduce the dimension of the data. The contribution of proteins to the identified principal components will be evaluated and the proteins with high contribution to the principal components are important and have strong influence to the distribution of the data. In addition, the data from each sample can be visualized in a scatter plot of any two principal components to reflect the relative variation of the multidimensional data. The samples sharing similar characteristics are expected to be located near each other in the scatter plots. Therefore, the principal component can be used to separate the data into subgroups when available and identify the subsets of samples that may be associated with different phenotypes under study. Note that the PCA usually cannot accurately define clear boundaries between different clusters or subsets in the data. The combined use of the PCA with clustering methods can help us to understand the cluster size, integrity and distribution.

12.5.6 Protein Networks

Although clustering provides important information on the dependencies between proteins, it fails to indicate the direction of the interaction between proteins, or whether the two proteins are just indirectly coregulated by common regulatory proteins. To provide a more refined information on the dependencies between proteins, graphical models have been proposed to extract a graphical representation of interacting proteins. The constructed graphical representation is a network of connected proteins, with nodes denoting the proteins and directed or undirected edges denoting the interaction between proteins.

Relevance networks (RNs) [39] are one of the simplest graphical models for constructing a protein network. The RNs are constructed based on pairwise association scores (ie, correlation coefficient) between proteins. The association scores between each pair of proteins are calculated and the pairs with a score exceeding some prespecified threshold value are connected by an undirected edge in the graph. Graphical Gaussian models (GGMs) [40] generalize the idea of the RN to use the partial correlation coefficient as the pairwise association score between proteins. The partial correlation [41] between the paired proteins measures their correlation conditional on all other proteins, so if these two proteins share large partial correlation, these two proteins are expected to have direct interactions. A Bayesian network (BN) [42] is a more sophisticated graphical model approach than the RN and GGMs approaches to construct the directed acyclic graphs (DAGs) to evaluate both the interaction between proteins and the direction of the protein–protein interaction. Specifically, the posterior probabilities for DAG will be statistically modeled, and the DAGs with highest posterior probabilities (scores) will be expected to describe the data well. More detailed description on the construction of BN can be found in Friedman et al. [42].

12.6 Summary

MS-based proteomics studies have been increasingly used for quantification of protein expression levels between biological conditions. When the proteomics experiment is conducted properly and the appropriate analytical method is used for data inference, MS-based proteomics experiments can provide great insight on the biological system.

However the experimental design and data analysis is a complex procedure given the variety and complexity of the experimental procedure. We have reviewed the critical issues and the statistical methods for both design and data analysis of proteomics data. Due to the complexity, it is important to involve a statistician at the early stage of the experimental planning prior to sample collection and data acquisition.

References

[1] Chahrour O, Cobice D, Malone J. Stable isotope labelling methods in mass spectrometry-based quantitative proteomics. J Pharm Biomed Anal 2015;113.

[2] Ong SE, Blagoev B, Kratchmarova I, Kristensen DB, Steen H, Pandey A, et al. Stable isotope labeling by amino acids in cell culture, SILAC, as a simple and accurate approach to expression proteomics. Mol Cell Proteomics 2002;1(5):376−86.

[3] Gygi SP, Rist B, Gerber SA, Turecek F, Gelb MH, Aebersold R. Quantitative analysis of complex protein mixtures using isotope-coded affinity tags. Nat Biotechnol 1999;17(10):994−9.

[4] Ross PL, Huang YN, Marchese JN, Williamson B, Parker K, Hattan S, et al. Multiplexed protein quantitation in *Saccharomyces cerevisiae* using amine-reactive isobaric tagging reagents. Mol Cell Proteomics 2004;3(12):1154−69.

[5] Montgomery DC. Design and Analysis of Experiments. 7th ed. Wiley; 2008.

[6] Kerr KF, Serikawa KA, Wei C, Peters MA, Bumgarner RE. OMICS 2007;11(2):152−65.

[7] Dobbin K, Simon R. Sample size determination in microarray experiments for class comparison and prognostic classification. Biostatistics 2005;6:27−38.

[8] Woo Y, Krueger W, Kaur A, Churchill G. Experimental design for three-color and four-color gene expression microarrays. Bioinformatics 2005;1:459−67.

[9] Kerr MK, Churchill GA. Experimental design for gene expression microarrays. Biostatistics 2001;2(2):183−201.

[10] Campbell MJ, Julious SA, Altman DG. Estimating sample sizes for binary, ordered categorical and continuous outcomes in two group comparisons. BMJ 1995;311:1145.

[11] Tibshirani R. A simple method for assessing sample sizes in microarray experiments. BMC Bioinformatics 2006;7:106.

[12] Cairns DA, Barrett JH, Billingham LJ, Stanley AJ, Xinarianos G, Field JK, et al. Sample size determination in clinical proteomic profiling experiments using mass spectrometry for class comparison. Proteomics 2009;9(1):74−86.

[13] Banjamini Y, Hochberg Y. Controlling the false discovery rate: a practical and powerful approach to multiple testing. J R Stat Soc Ser B 1995;57:289—300.

[14] Oberg AL, Vitek O. Statistical design of quantitative mass spectrometry-based proteomic experiments. J Proteome Res 2009;8(5):2144—56.

[15] Oberg AL, Mahoney DW. Statistical methods for quantitative mass spectrometry proteomic experiments with labeling. BMC Bioinformatics 2012;13(Suppl. 16): S7-2105-13-S16-S7. Epub November 5, 2015.

[16] Shilov IV, Seymour SL, Patel AA, Loboda A, Tang WH, Keating SP, et al. The Paragon Algorithm, a next generation search engine that uses sequence temperature values and feature probabilities to identify peptides from tandem mass spectra. Mol Cell Proteomics 2007;6(9):1638—55.

[17] Cox J, Mann M. MaxQuant enables high peptide identification rates, individualized p.p.b.-range mass accuracies and proteome-wide protein quantification. Nat Biotechnol 2008;26:1367—72.

[18] Perez-Riverol Y, Wang R, Hermjakob H, Müller M, Vesada V, Vizcaíno JA. Open source libraries and frameworks for mass spectrometry based proteomics: a developer's perspective. BBA Proteins Proteom 2014;1844(1):63—76.

[19] Oberg AL, Mahoney DW, Eckek-Passow JE, Malone CJ, Wolfinger RD, Hill EG, et al. Statistical analysis of relative labeled mass spectrometry data from complex samples using ANOVA. J Proteome Res 2008;7(1):225—33.

[20] Mahoney DW. http://pubs.acs.org/doi/suppl/10.1021/pr700734f/suppl_file/pr700734ffile001.pdf; 2009.

[21] Karpievith Y, Stanley J, Taverner T, Huang J, Adkins JN, Ansong C, et al. A statistical framework for protein quantitation in bottom-up MS-based proteomics. Bioinformatics 2009;25(16):2028—34.

[22] Luo R, Colangelo CM, Sessa WC, Zhao H. Bayesian analysis of iTRAQ data with nonrandom missingness: identification of differentially expressed proteins. Stat Biosci 2010;1(2):228—45.

[23] Webb-Robertson BM, Wiberg HK, Matzke MM, Brown JN, Wang J, McDermott JE, et al. Review, evaluation, and discussion of the challenges of missing value imputation for mass spectrometry-based label-free global proteomics. J Proteome Res 2015;14(5):1993—2001.

[24] Wiederin JL, Donahoe RM, Anderson JR, Yu F, Fox HS, Gendelman HE, et al. Plasma proteomic analysis of simian immunodeficiency virus infection in rhesus macaques. J Proteome Res 2010;9(9):4721—31.

[25] Kraft-Terry SD, Engebresten LL, Bastola DK, Fox HS, Ciborowski P, Gendelman HE. Pulsed stable isotope labeling of amino acids in cell culture uncovers the dynamic interactions between HIV-1 and the monocyte-derived macrophage. J Proteome Res 2001;10(6):2852—62.

[26] Margolin AA, Ong S-E, Schenone M, Gould R, Schreiber SL, Carr SA, et al. Empirical Bayes analysis of quantitative proteomics experiments. PLoS One 2009;4(10):e7454.

[27] Schwacke JH, Hill EG, Krug EL, Comte-Walters S, Schey KL. iQuantitator: a tool for protein expression inference using iTRAQ. BMC Bioinformatics 2009;10:342.

[28] Jow H, Boys RJ, Wilkinson DJ. Bayesian identification of protein differential expression in multi-group isobaric labelled mass spectrometry data. Stat Appl Genet Mol Biol 2014;13(5):531–51.

[29] Westfall PH, Young SS. Resampling based multiple testing: examples and methods for p-value adjustment. New York: Wiley; 1993.

[30] Storey JD, Tibshirani R. Statistical significance for genomewide studies. Proc Natl Acad Sci USA 2003;100(16):9440–5.

[31] Efron B. Large scale simultaneous hypothesis testing: the choice of a null hypothesis. J Am Stat Assoc 2004;99:96–104.

[32] Efron B. Local false discovery rates. Technical Report. Stanford (CA): Department of Statistics, Stanford University; 2005.

[33] Efron B. Correlation and large scale simultaneous significance testing. Technical Report. Stanford (CA): Department of Statistics, Stanford University; 2006a.

[34] Efron B. Size, power, and false discovery rates. Technical Report. Stanford (CA): Department of Statistics, Stanford University; 2006b.

[35] Storey JD. The positive false discovery rate: a Bayesian interpretation and the q-value. Ann Stat 2003;31:2013–35.

[36] Pounds S, Morris SW. Estimating the occurrence of false positives and false negatives in microarray studies by approximating and partitioning the empirical distribution of p-values. Bioinformatics 2003;19(10):1236–42.

[37] Eisen MB, Spellman PT, Brown PO, Botstein D. Cluster analysis and display of genome-wide expression patterns. Proc Natl Acad Sci USA 1998;95(25):14863–8.

[38] Joliffe T. Principal components analysis. Berlin: Springer; 1986.

[39] Butte AS, Kohane IS. Mutual information relevance networks: functional genomics clustering using pairwise entropy measurements. Pac Symp Biocomput 2000;5:418–29.

[40] Edwards D. Introduction to graphical modelling. 2nd ed. New York: Springer Verlag; 2000.

[41] Shipley B. Cause and correlation in biology: a user's guide to path analysis, structural equations and causal inference. Cambridge University Press; 2002.

[42] Friedman N, Linial M, Nachman I, Pe'er D. Using Bayesian networks to analyze expression data. J Comput Biol 2000;7:601–20.

13

PRINCIPLES OF ANALYTICAL VALIDATION

J. McMillan

University of Nebraska Medical Center, Omaha, NE, United States

CHAPTER OUTLINE

13.1 Introduction   239
13.2 Liquid Chromatographic Methods   240
13.3 Validation of a Liquid Chromatographic Method: Identity, Assay, Impurities   241
13.4 Recovery   244
13.5 Accuracy   244
13.6 Precision   245
13.7 Calibration Curve, Linearity, and Sensitivity   245
13.8 Selectivity and Specificity   246
13.9 Stability   247
13.10 Aberrant Results and Errors in Analyses   247
13.11 Further Development of Methods Validation   249
References   250

13.1 Introduction

Analytical validation can be defined as the collection and evaluation of data generated from the process/method used in making a product whether it is commercial, experimental or a scientific study. Analytical validation establishes experimental evidence that a process/method/study consistently delivers reproducible, precise and accurate results using established and accepted methodology. Analytical validation consists of multiple steps and starts with a validation master plan. A validation master plan has broad scope and will contain more

Proteomic Profiling and Analytical Chemistry. http://dx.doi.org/10.1016/B978-0-444-63688-1.00013-6

elements with higher stringency for validation of a commercial product than one for validation of results of an experiment that is aimed at publication. A validation master plan, although not absolutely necessary in a laboratory experimental setting, is very helpful for downstream discovery data presentation and should clarify general objectives, procedures and protocols and prioritize validation steps. It should include a description of the equipment to be used with specific parameters such as dynamic range of signal measurement, volumes of samples to be measured, etc. Although not every laboratory validation procedure requires all these principles, all should be considered while planning the validation process. They are specificity, linearity robustness, range, detection limit, quantitation limit, ruggedness, selectivity and sustainability. Here we discuss those that are important in validation of liquid chromatography, a technique that is an integral part of every proteomics study.

13.2 Liquid Chromatographic Methods

Chromatography has become a mainstay of separation technology for research, the pharmaceutical industry and clinical analysis. Liquid chromatography coupled to mass spectrometry (LC/MS) is now the method of choice of proteomics for protein identification [1,2]. Separation of biological compounds by high-performance liquid chromatography (HPLC) has been used for decades by the pharmaceutical and biotechnology industries for resolution, reproducibility and sensitivity for compounds that cannot be analyzed by gas chromatography [3]. The broad selection of mobile phases and stationary phases used in HPLC has also made it an important tool for proteomics [2,3]. Conventional HPLC operates with column and pump pressures up to 3600 and 6500 psi, respectively, and stationary-phase particle sizes in the range of 3–5 μm. To increase sensitivity for proteomics analyses, column internal diameters (IDs) have been greatly reduced (from the conventional 4.6 mm ID to 75 μm ID for nano-HPLC columns) and new stationary phases have been developed [4]. HPLC using nano-bored columns packed with newer stationary phases has provided good separation of relatively complex mixtures. However, higher resolution is required for the thousands of peptides encountered in proteomics analyses. Dramatic improvements in resolution, sensitivity and separation speed have occurred with the development of ultra-high-pressure liquid chromatography (UPLC or UHPLC) [3,5]. UPLC uses smaller column particle size (<2 μm) and higher pressure (15,000 psi) to reduce analysis time and improve sensitivity and resolution [6]. To operate under such high pressure and in a pH range of $1-12$, UPLC columns are composed of an ethylene-bridged hybrid (BEH) particle structure that provides mechanical stability to the packing material. The increase in sensitivity and resolution afforded by UPLC is illustrated in Fig. 13.1 [5]. In this illustration, the same mouse urine sample was analyzed using HPLC-MS (Fig. 13.1A) and UPLC-MS (Fig. 13.1B), where increased peak resolution and improvements in peak shape and sharpness can clearly be seen with UPLC separation. As a consequence of the increase in peak resolution and decrease in run time a lesser amount of solvent is used and sample throughput is increased. To determine whether a new analytical LC technique or method (nanoHPLC or UPLC versus HPLC) provides quantifiable analytical advantages, it is important to directly compare the new and old techniques under the same conditions. However, regardless of the separation method being used for analysis, proper validation of a new method or change in method is required. The remainder of this chapter will discuss the components of method validation.

13.3 Validation of a Liquid Chromatographic Method: Identity, Assay, Impurities

When developing a bioanalytical method it is important to demonstrate that it is accurate and reproducible over the required range of concentrations for the analyte in a particular biological matrix.

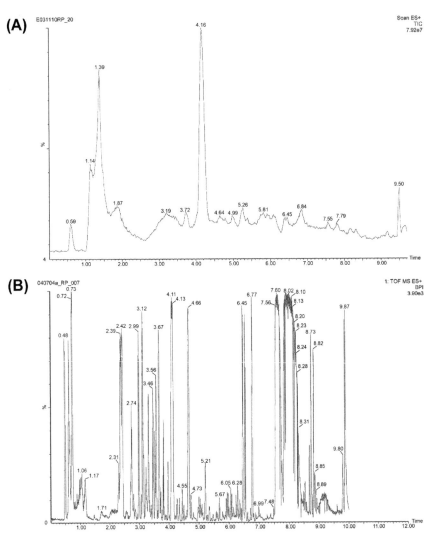

Figure 13.1 Chromatograms obtained by analyzing the same sample with HPLC (A) and UPLC (B). Chromatographic separation of white female AM mouse urine using (A) 2.1 × 100 mm Waters Symmetry 3.5-μm C18 column, eluted with 0—95% linear gradient of water with 0.1% formic acid: acetonitrile with 0.1% formic acid over 10 min at a flow rate of 0.6 ml/min. The column eluent was monitored by ESI oa-TOF-MS from 50 to 850 *m/z* in positive ion mode. (B) Sample analyzed using a 2.1 × 100 mm Waters ACQUITY 1.7-μm C18 column and eluted with a linear gradient of 0—95% water with 0.1% formic acid: acetonitrile with 0.1% formic acid over 10 min at a flow rate of 0.5 ml/min. The column eluent was monitored by ESI oa-TOF-MS from 50 to 850 *m/z* in positive ion mode. Reprinted with permission from Wilson ID, Nicholson JK, Castro-Perez J, Granger JH, Johnson KA, Smith BW, et al. High resolution "ultra performance" liquid chromatography coupled to oa-TOF mass spectrometry as a tool for differential metabolic pathway profiling in functional genomic studies. J Proteome Res 2005;4:591—8. Copyright 2005 American Chemical Society.

For development and validation of bioanalytical chromatographic separation methods, there are recommendations from several national and international organizations to ensure that the data provided for marketing and clinical applications are uniformly acquired [7–9]. However, there is no single final guideline for method validation. It is important to understand that the degree or extent of studies needed for method validation depends on the purpose of the validation. Thus the first step in method validation is to define the objective of the method. A quantitative method (patient monitoring, final product potency, level of impurities and contaminants) may require more validation steps than a qualitative method for component identity. Full method validation is required when a new method is developed or when additional analytes are added to an already existing assay. Partial validation is a modification to an already accepted method when full validation is not needed. Method changes that would be included in this category are transfer of a method between laboratories, use of new hardware or software for data acquisition, changes in bioanalytical matrices, demonstration of analyte in the presence of specific metabolites, demonstration of analyte in the presence of concomitant treatments or additional contaminants, a change in the analytical method parameters, and a change in the processing of samples or use of rare matrices [9]. Cross-validation compares the results obtained from two different analytical methods, and is required when data from the same study is analyzed by two or more methods.

The establishment of a validated method is based upon the parameters of accuracy, precision, selectivity, sensitivity, reproducibility and stability. The guidelines from various national and international regulatory agencies are in general agreement over the requirements for these parameters [7,10]. For proteomics methods the number of analytes is huge and work on all analytes may be impractical; thus a small number of analytes (eg, 10) over a range of molecular weights and hydrophobicities may be used for establishment and validation of the method [11]. Once a method is established and validated, a detailed description of the method should be prepared in the form of a Standard Operating Protocol. The general guidelines for these parameters will be discussed in the next sections.

13.4 Recovery

Recovery is determined by comparing the detector response of a known quantity of analyte added to and extracted from the biological matrix and the response of the same concentration of pure analyte in vehicle or mobile phase. Extraction of an analyte from the sample matrix, ie, extraction efficiency, should be well-characterized using a range of spiking concentrations [12]. For HPLC, by including an internal standard in the extraction solvent recovery efficiency can be determined. When an internal standard is used, its recovery should be similar to that of the analyte and should be reproducible and it should give a reliable response. The amount of internal standard should be well above the limit of quantitation but should not suppress the response of the analyte. Determination of recovery of the analyte and internal standard at low, medium, and high concentrations is recommended by the *Journal of Chromatography B* guidelines [13].

13.5 Accuracy

Accuracy is the closeness of the value obtained by the analytical method to the true value. To determine accuracy, replicate analysis of samples is done by (1) comparing the values obtained with known replicate samples to the true value, (2) comparing results of the new method to those obtained with another established method, (3) spiking the analyte into different matrices and (4) use of standard additions when it is not possible to obtain matrices without the presence of analyte [8]. The FDA recommends a minimum of three concentrations in the range of expected concentrations and a minimum of five determinations per concentration [10]. Deviations from the expected value should be no more than 15% for all concentrations except the lower limit of quantitation (LLOQ), where 20% is accepted. The measure of accuracy is thus the deviation of the mean of the actual value from the true value.

13.6 Precision

In conjunction with accuracy, precision (coefficient of variation) of the method is determined to describe the degree of repeatability of the method under normal operations. Three levels of precision are described by documents from the International Conference on Harmonization (Q2R1) and the International Organization for Standardization [7]. The first is repeatability, ie, the precision of the assay over short periods of time. The second is intermediate precision, which refers to the variations in results within a laboratory that occur on different days, with different analysts or with different equipment. The third is reproducibility and reflects the differences in assay results between laboratories. Precision is determined using a minimum of five determinations per concentration at a minimum of three concentrations in the expected range. It can be expressed as the percent coefficient of variation (%CV) of the replicate measurements (%CV = standard deviation/mean × 100). Variation at each concentration level should not exceed 15% of the CV with 20% acceptable at the LLOQ. Graphing intra-day and inter-day values for the internal standard can aid in monitoring assay precision; variation of an internal standard value that exceeds 2 standard deviations from the mean may indicate technical problems with the analysis that need to be addressed.

13.7 Calibration Curve, Linearity, and Sensitivity

The calibration curve defines the relationship between the detector response and the concentration of analyte in the sample matrix. For multiple analytes, a sample calibration curve is generated for each analyte. To fit the standard curve, the simplest algorithm that describes the concentration/response relationship is used. Thus the algorithm may be linear or nonlinear [7,9] but should minimize percent relative error. In the case of liquid chromatography with tandem mass spectrometry detection, forcing the data to a linear function may result in large errors in measurements of results. The calibration curve should consist of five to eight points that cover the entire range of expected analyte concentrations in the test samples, ie, from 0 to 200% of the theoretical content. The lowest concentration should be the LLOQ and the highest concentration should be the upper limit of quantitation (ULOQ). If sample analyte results fall outside the range of the LLOQ or ULOQ, the sample should be diluted in matrix and a new standard curve in matrix prepared.

The LLOQ is the lowest concentration that can be defined with accuracy and precision. To define the LLOQ at least five samples independent of standards should be used and the CV or confidence interval determined. Conditions to define the LLOQ include a response at least five times that of the blank response and peak accuracy of 80–120% and precision of 20%. For chromatographic methods, the LLOQ is based on the signal-to-noise ratio (S/N). Signal and baseline noise is defined as the height of the analyte peak (signal) and the amplitude between the highest and lowest point of the baseline noise in the area around the analyte peak. The S/N for LLOQ is usually required to be ≥10 [13]. The LLOQ is not the limit of detection, which is the lowest concentration that the method can reliably differentiate from background noise.

13.8 Selectivity and Specificity

When evaluating a method a key criterion is the ability of the method to differentiate analyte from other sample components (contaminants, matrix components, degradation products, etc.). To determine selectivity, the quantitation of analyte in test matrices containing all potential components is compared to quantitation of analyte in solution alone. The specificity of the assay determines that the obtained signal is due to the analyte of interest and that there is no interference from other matrix components, impurities or degradation products. Peak shape when used in conjunction with diode array, MS, or MS/MS detection can be used to determine the purity of a peak [13].

13.9 Stability

The stability of the analyte in the biological matrix under a variety of conditions pertinent to collection, storage and analysis should be determined, including stability in stock solutions. First, stability of the analyte over three freeze/thaw cycles at two concentrations is recommended. Second, the stability of three aliquots of sample at room temperature for up to 24 h, ie, based upon the period of time the samples would remain at room temperature during the study, should be determined. Third, the stability of the samples under expected storage conditions for a period of time exceeding the projected time of the study should be determined for three aliquots at two concentrations. Fourth, stock solution and internal standard stability should be determined at room temperature over a period of 24 h and at the expected storage conditions for the period of the study. Fifth, once the samples have been processed for analysis, the stability of the samples during the period of analysis should be determined. This includes stability of the analyte and internal standard under conditions that replicate that of the autosampler during analysis. Stability tests are performed against freshly prepared analyte standards analyzed in the same run. Changes in stability of $\leq 10\%$ are generally acceptable. If instability of the samples or standards is observed, use of buffers, antioxidants, enzyme inhibitors, etc., may be necessary to preserve the integrity of the analytes.

13.10 Aberrant Results and Errors in Analyses

Before beginning an analytical method, the suitability of the system to deliver reliable and repeatable results should be determined. Parameters that can be evaluated and compared to expected results include plate count, tailing, peak resolution and repeatability (retention time and peak area). When results are obtained that are outside of the acceptable range defined for the method, the cause of the aberration should be investigated. The investigation should systematically determine whether the aberrant result is due to malfunctioning equipment, an error in sample preparation or analysis or an error in sample collection. Quality control (QC) standards of various concentrations should be interspersed with samples during a test run. At least 67% (four of six) of the QC samples should fall within 15% of their expected values; 33% of QC values may fall outside of the 15% of expected values, but they should not all be replicates of a single concentration [13]. An erroneous result for QC samples might suggest a malfunction in the HPLC system or detector. If the equipment is functioning within previously set specifications, then an investigation of the preparation and analysis of the sample is warranted. A first check should confirm that the calculations used to convert raw data into final result were correct. In addition, it is recommended that a check for usage of proper standards, solvents, reagents, and other solutions be performed. To determine whether the samples were prepared properly or the aberrant result might be due to an equipment malfunction, reinjection of the samples is possible. Reanalysis of the original sample will determine whether the sample itself is different or the sample was processed incorrectly, eg, improper dilution, incomplete extraction, inadequate resuspension of dried samples, etc. To determine whether an extraction was carried out to completion, reextraction of a sample can be done. However, if it is found that the sample was not fully extracted, a reevaluation and revalidation of the method should be performed using the modified extraction protocol.

Once new results are obtained, how should the information be reconciled with the initial aberrant result? Two methods that are recommended by FDA Guidelines include averaging and outlier results [14]. First, averaging can be an appropriate approach, but its use depends on the purpose of the sample, the type of assay being performed, and whether the sample is homogeneous. For HPLC results, peak responses can be averaged from consecutive injections of the same sample and the average of the peak's responses would be considered the response for that sample. Analysis of different portions from the original sample would be done to determine the variability/homogeneity of the original sample. The cause of unusual variations in replicate sampling should be investigated. Averaging can, however, conceal variations in individual test results that might indicate non-homogeneity of the original sample. Thus, it is inappropriate to use average results if the purpose of the analytical test is to determine sample variability.

Second, values that are significantly different from others in a series of replicate measurements may be statistical outliers. A deviation in response may be due to an error in the analytical method or due to an inherent variability in the tested sample. To determine the relevance of extreme results, a statistical procedure for determining outlier values may be used. If a result is determined to be a statistical outlier, the cause of the aberrant response should be investigated. As with averaging, if the purpose of the analysis is to determine homogeneity of a sample, an outlier test should not be used.

13.11 Further Development of Methods Validation

The purpose of method validation is to demonstrate acceptability of a method for a particular analysis. With the continued development of higher-resolution HPLC instrumentation and detection systems, such as higher-sensitivity mass spectrometry and tandem mass spectrometry systems, and improved software for analysis, there is a need to determine the robustness and reproducibility of data obtained from these improvements [15]. For statistical validation of proteomics methods, at least three technical replicates (same sample analyzed three times) and three biological replicates (eg, tissue from three animals in the same group) should be analyzed. By taking a stepwise logical approach to method validation, it can be demonstrated to scientific peers, regulatory agencies and potential business partners that the method will produce reliable, believable results.

References

[1] Ayrton J, Dear GJ, Leavens WJ, Mallett DN, Plumb RS. Optimisation and routine use of generic ultra-high flow-rate liquid chromatography with mass spectrometric detection for the direct on-line analysis of pharmaceuticals in plasma. J Chromatogr A 1998;828:199−207.

[2] Wolters DA, Washburn MP, Yates 3rd JR. An automated multidimensional protein identification technology for shotgun proteomics. Anal Chem 2001;73:5683−90.

[3] Shi Y, Xiang R, Horvath C, Wilkins JA. The role of liquid chromatography in proteomics. J Chromatogr A 2004;1053:27−36.

[4] Mitulovic G, Mechtler K. HPLC techniques for proteomics analysis−a short overview of latest developments. Brief Funct Genomic Proteomic 2006;5:249−60.

[5] Wilson ID, Nicholson JK, Castro-Perez J, Granger JH, Johnson KA, Smith BW, et al. High resolution "ultra performance" liquid chromatography coupled to oa-TOF mass spectrometry as a tool for differential metabolic pathway profiling in functional genomic studies. J Proteome Res 2005;4:591−8.

[6] Yandamuri N, Nagabattula KRS, Kurra SS, Batthula S, Allada LPS, Bandam P. Comparative study of new trends in HPLC: a review. Int J Pharm Sci Rev Res 2013;23:52−7.

[7] Rozet E, Ceccato A, Hubert C, Ziemons E, Oprean R, Rudaz S, et al. Analysis of recent pharmaceutical regulatory documents on analytical method validation. J Chromatogr A 2007;1158:111−25.

[8] Shabir GA. Validation of high-performance liquid chromatography methods for pharmaceutical analysis. Understanding the differences and similarities between validation requirements of the US Food and Drug Administration, the US Pharmacopeia and the International Conference on Harmonization. J Chromatogr A 2003;987:57−66.

[9] Shah VP, Midha KK, Findlay JW, Hill HM, Hulse JD, McGilveray IJ, et al. Bioanalytical method validation−a revisit with a decade of progress. Pharm Res 2000;17:1551−7.

[10] FDA US. Guidance for industry. bioanalytical method validation. September 2013. Draft Guidance, http://www.fda. gov/downloads/drugs/guidancecomplianceregulatoryinformation/ guidances/ucm368107.pdf.

[11] Krull I, Kissinger PT, Swartz M. Analytical method validation in proteomics and peptidomics studies. 2008. http://www.chromatographyonline.com/analytical-method-validation-proteomics-and-peptidomics-studies [accessed 12.08.15].

[12] Carr GP, Wahlich JC. A practical approach to method validation in pharmaceutical analysis. J Pharm Biomed Anal 1990;8:613−8.

[13] Tiwari G, Tiwari R. Bioanalytical method validation: an updated review. Pharm Methods 2010;1:25−38.

[14] FDA US. Guidance for Industry. Investigating Out-of-Specification (OOS) test results for pharmaceutical production. October 2006. http://www.fda.gov/downloads/Drugs/.../Guidances/ucm070287.pdf.

[15] Gorog S. The changing face of pharmaceutical analysis. Trends Anal Chem 2007;26:12−7.

VALIDATION IN PROTEOMICS AND REGULATORY AFFAIRS

J. Silberring
*AGH University of Science and Technology, Krakow, Poland;
Polish Academy of Sciences, Zabrze, Poland*

M. Wojtkiewicz and P. Ciborowski
*University of Nebraska Medical Center, Omaha, NE, United
States*

CHAPTER OUTLINE
14.1 The "Uphill Battle" of Validation   253
14.2 Accuracy and Precision   256
14.3 Experimental Design and Validation   258
14.4 Validation of the Method   259
14.5 Validation of Detection Levels   260
14.6 Validation of Reproducibility and Sample Loss   262
14.7 Validation of Performance of Instruments   263
14.8 Bioinformatics: Validation of Output of Proteomic
    Data   266
14.9 Cross-Validation of Initial Results   267
14.10 Proteomics and Regulatory Affairs   267
References   269

14.1 The "Uphill Battle" of Validation

Experimental design principles for achieving validity and efficiency are required for any experiment that has one or more variables, whether inherent or introduced by the investigator [1,2]. Traditionally, such principles are recognized for low-throughput experiments, but have become accepted for

Proteomic Profiling and Analytical Chemistry. http://dx.doi.org/10.1016/B978-0-444-63688-1.00014-8

high-throughput procedures, such as microarray experiments [3]. In this chapter we attempt to review the validation principles and their applicability to complex high-throughput experiments such as proteomic profiling. As presented in chapter 1, it has to be recognized that the typical proteomic experiment/study consists of a string of multiple steps of which not all are represented by typical analytical procedures. Nevertheless, all steps will have impact on the overall outcome, providing answers of biological importance. As much as validation of analytical methods is quite well established (see chapter: Principles of Analytical Validation for details), validation of experimental design or bioinformatics analysis is still in infancy.

There is no "one size fits all" in validation, or in a multistep proteomic profiling experiment in particular, when we try to increase the sensitivity of every step of the entire proteomic study [4]. Furthermore, even if every step can be validated separately, it does not necessarily translate into being able to validate the final outcome by orthogonal method(s). This is because of three major reasons: (1) each step is governed by specific analytical parameters that are different than the entire process in question; (2) biological processes are dynamically changing over time (often quickly) and at multiple levels; and (3) in many if not most instances, we are not able to define the relationship between rate of change and biological effect. A plot of fold change in a protein's biological activity versus overall change in function of the studied system would be very helpful in validation; however, this is usually the very question we ask and try to answer using profiling experiments. This subsequently deprives us from points of reference which are critical for validation [5,6]. Studying changes in proteomes of humans is even more complicated, not only because of the complexity of the human organism but also because ethical boundaries limit how far we can manipulate this system. Animal models that are very valuable in reductionist studies are less informative about functions of a human body in its entirety in holistic studies.

Validation procedures are time-consuming and not as spectacular as thousands of identified compounds. Therefore, validation and internal laboratory quality control, which is a mandatory routine in analytical chemistry, needs to be transferred and adapted to proteomic experiments which, as we said above, are much more complicated. Although, we are usually interested in validation of the final output, any given methodology in the multistep procedure is a subject of validation. The common terms, such as accuracy, precision, specificity, linearity, can be found in any book on analytical chemistry or medicinal chemistry. Similarly, detailed guidelines for testing those parameters and valuable advice can be found online, eg, posted by the International Union of Pure and Applied Chemistry (IUPAC, http://www.iupac.org/) or the European Medicine Agency (EMA, http://www.ema.europa.eu). The American Association of Pharmaceutical Scientists (AAPS, http://www.aaps.org), Food and Drug Administration (FDA, http://www.fda.gov), and many other international and national agencies prepare their own documents and recommendations. These publications are devoted to standardized analytical procedures to maintain unified safety of drugs, detection of impurities, fulfilling goals to maintain and procedures for validation and control of various products. In the field of proteomics, infrastructure for unified validation procedures is not as well developed and/or structured as in the pharmaceutical industry, environmental analyses, forensics, etc. We would like to acknowledge efforts of organizations such as the Human Proteome Organization (HUPO; http://www.hupo.org/) and Association of Biomolecular Research Facilities (ABRF, http://www.abrf.org/) for evaluating collected results, organizing various initiatives to foster and coordinate novel technologies, disseminating knowledge, performing statistical evaluation of collaborative trials, providing certified standards, etc.

In this chapter we will discuss issues related to an "uphill battle" of validation of each step of proteomic study as depicted in Fig. 14.1. At the end we will briefly discuss how we may need to approach proteomic validation from perspective of regulatory affairs. This is an emerging problem as transgenic organisms are

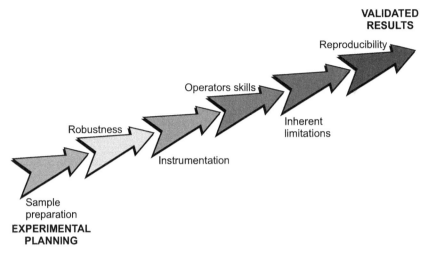

Figure 14.1 An uphill battle of validation. Validation has to consider multiple steps, and if processes are not properly validated at any of these steps, the final product, eg, biomarker will not be successfully validated. On the other hand, if each step is successfully validated, the entire process might not pass the overall validation.

more and more often used for mass production of various products including food sources.

14.2 Accuracy and Precision

The recent explosion of research based on experimentation across the entire world leads quite often to miscommunication resulting from differences, sometimes subtle, in understanding terminology. This is a critical issue for validation, which cannot accept anything that goes off-track in "speaking one precise language." As much as we can discuss which definition of, eg, systems biology reflects the most closely descriptive intensions of this term, validation must use definitions that cannot carry any doubt. Otherwise results will not be comparable at the required level. Here we bring up one example of the definitions of "accuracy" and "precision." Based on many definitions available, accuracy is the condition, quality or degree of closeness of the true, correct, or exact quantity to that quantity's actual (true) value. Subsequently, precision is exactness in measurement, execution to be consistently reproduced and strictly distinguished from others.

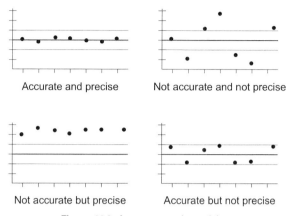

Figure 14.2 Accuracy and precision.

For both accuracy and precision, the number of significant digits to which a value has been reliably measured is very important if not critical. Precision also contains the degree of reproducibility and of repeated measurements that yield the same results under unchanged conditions. Fig. 14.2 shows graphically the meanings of accuracy and precision.

It is much easier to satisfy these two conditions when we are measuring static systems such as mixture of end-products of a chemical synthesis that is fully stopped. It is much more challenging when a dynamic and complex biological system is the subject of measurements, because it is extremely difficult to define "unchanged conditions." This has a profound implication for how researchers describe their experimental conditions and analytical steps.

Accuracy may deviate in any analysis due to systematic errors, such as improper calibration of instruments or constant mistakes of the operator [7]. Precision may depend on the operator skills, stability of instrumentation, etc. The sum of all these errors in parallel with variety of instrumentation and principles of technology platforms, ie, ion traps versus quadrupole time-of-flight, will have a major impact on the quality of the obtained proteomic set of data. Therefore, we have to expect that further validation of potential biomarkers in an independent test may give unexpected results. For more information in this area we direct readers to the International Vocabulary of

Metrology—Basic and General Concepts and Associated Terms (VIM) (http://www.bipm.org/utils/common/documents/jcgm/JCGM_200_2008.pdf).

14.3 Experimental Design and Validation

Proteomic profiling, like other experiments, has to be designed in such a way that when executed, factors that cause or contribute to variability are being properly controlled and output falls within the limits of the ranges that are up-front accepted. This implies that proteomic experiments should be performed based on as many possible criteria that had been already established. That said, we expect that biological systems when manipulated consistently and reproducibly behave as expected. For example, if we stimulate macrophage with lipopolysaccharide (LPS), we expect that cells start secreting the expected cytokines [8]. Therefore, experimental design of proteomic experiments has to define the system empirically. For the purpose of our considerations here, we will define validation of experimental design as an establishment of evidence which provides a high confidence that the biological system being investigated will produce an outcome consistent with predetermined parameters. Such a goal is gradually much harder to be achieved along with increasing complexity of the biological system and much more sophisticated experimental schemes even supported by the newest technologies. For example, the response to any given stimuli of transformed cells under defined culture conditions will be much more uniform than the response to the same stimulus of primary cells isolated from various human subjects or even inbred animals [9]. The situation is further complicated when samples such as plasma/serum or CSF represent a complex biological system as a snapshot at one time point, thus reflecting only this point of dynamic physiological state of control and diseased subjects [10]. As much as plasma/serum or CSF is in a way a "homogenous" sample in the sense that it consists of a mixture of proteins (after removing all metabolites), tissue biopsy is a mixture of different types of cells. In this situation, establishing criteria empirically that define such a system is very, if not extremely, difficult and inevitably many parameters might be overlooked, leading to aberrations in validation. This requires procedures that need to be established for monitoring the output and validation of the performance of those factors that might be a source of variability.

The initial question we should ask when designing proteomic experiments is whether full unbiased or targeted proteomic profiling will better serve in testing our hypothesis. As much as such a question seems straightforward, in reality it is not and many factors need to be considered. The first and foremost factor is whether our biological system secures a sufficient amount of biological material to perform replicate analyses for validation using orthogonal methods. At this point we must consider how we will approach validation of our overall results when the experiment is executed. How much biological material needs to be saved for validation purposes, and at which state of sample processing? The high dynamic range of protein concentrations in CSF/plasma/serum requires an initial step of fractionation (ie, immunodepletion), starting with albumin, which has a wide range of concentration within the patient population and is a major source of variability. One big question in validation of plasma/serum/CSF biomarkers is whether changes in levels of any given protein should be validated in body fluid as initially used as a sample source or validation should be performed on samples after immunodepletion of the most abundant proteins. It has been shown that downstream orthogonal validation using pre- and postprocessed samples usually does not match, thus requiring novel approaches in biomarker validation.

14.4 Validation of the Method

According to an IUPAC definition (M. Thompson et al. © 2002 IUPAC, Pure and Applied Chemistry 74, 835–855), validation applies to a defined protocol, for the determination of a specified analyte and range of concentrations in a particular type of tested material, used for a specified purpose. The procedure should check that the method performs adequately for the purpose throughout the range of analyte concentrations and test materials to which it is applied. In general, such a definition of validation can be implemented in analytical chemistry where a strictly defined method is to be concerned. Proteomics strategies, however, deal with biological samples that undergo complex extraction and fractionation prior to measurements [11,12]. This effect of low reproducibility of procedures is well known in biochemistry when the reductionist approach is used, and must be considered for -omics approaches as well. The classical example might be chromatographic separation of the same material, aliquoted and frozen in several portions. Separation of such material in a certain time-frame would never produce identical profiles. For instance, a tissue sample obtained from one patient during, eg, surgery may significantly differ from another sample due to, eg, another team of surgeons, the patient's diet, pharmacotherapy, etc. This is often referred to as sampling uncertainty. From this point of view, it is more relevant to refer to the "analytical system" rather than the "analytical method." Luckily, nowadays proteomics compares profiles from several samples simultaneously, which at least unifies part of the methodology. It must be clearly stated here that complete consistency and thus standardization of sample withdrawal would remain the major obstacle in further validation and quality control of proteomic strategy and will have a significant impact on the robustness of the method. The major drawback of "-omics" methods is that they are considered nonroutine. It is not common that the entire workflow is identical for each sample type (eg, plasma/serum, tissue, cell culture, etc.). For instance, body fluids require immunodepletion whereas tissues are processed using homogenization and/or organelle fractionation.

14.5 Validation of Detection Levels

The detection limit is a crucial factor in determining whether a molecule can be identified and quantified with an acceptable level of confidence. Thus the detection limit can be defined as the smallest amount or concentration of an analyte in the test sample that can be reliably distinguished from the baseline. If good and pure standards are available, protocols of validating detection limit are straightforward. To avoid influence of other compounds that are present in the sample, an addition of identical analyte with stable isotopes, such as ^{13}C and ^{15}N, appears the best approach [13]. Spiked-in "heavy" analyte will coelute during liquid chromatography separation (internal calibration) but will be recognized as a separate peak by the mass spectrometer as a distinct molecular species. This is opposite to external calibration of the detection level, which is comprised of a separate analytical run where a known amount of pure standard is used. Both methods are successfully applied for low-complexity samples containing a handful of compounds with similar analytical characteristics.

High-complexity samples that are subject to high-throughput profiling analyses pose additional challenges in validation of detection levels. Such samples contain thousands or tens of thousands of peptides with a wide range of analytical properties, making it impossible to create simple and reliable standards with applicability across such a broad spectrum of biochemical properties. One approach, although not quantitative per se, is to set a signal-to-noise (S/N) ratio threshold to define sufficient strength of a signal for quantitative comparisons. The S/N ratio is often used for MS^n experiments because it allows for comparisons of analytical runs. A S/N factor of 3:1 is quite often used as a threshold; however, for quantitation it should not be lower than 5:1 and even 10:1 for rigorous clinical assays.

Recently, Geiger and coworkers [14] proposed a mixture of cell lysates of five different cell lines labeled with "heavy" Arg and Lys to be mixed with lysate of unlabeled tissue. One can make an assumption that in this example each peptide from the tested sample will have its "heavy" counterpart. One caveat in this approach is that such standard is good as long as one pool of samples lasts, and thus one has a source of a standard. Subsequent mixture of five cell cultures may have different ratios, and considering the complexity of such an internal standard, it cannot be reproduced and/or normalized. Thus results from the experiment performed using batch 1 (pooled samples 1) cannot be fully compared to the results from experiments performed with batch 2, 3 or subsequent (pooled samples 2, 3 and subsequent) [14]. Alternative analysis would be employed, and is quite often, to use iTRAQ methodology when a control or reference sample has one reporter ion assigned.

Regardless of the strengths and weaknesses of each approach, validation of detection levels in complex samples should be considered in an early phase of proteomic experiment planning and must be considered during data analysis, in particular when precise quantitation plays a crucial role.

14.6 Validation of Reproducibility and Sample Loss

Inter- and intra-assay precision are two approaches to validating reproducibility. The intra-assay precision of a method is the measure of how the individual test results of multiple injections of a series of standards agree. This is characterized by standard deviation or standard error of mean. Precision should be calculated from at least 10 replicates of standards at various concentrations (low, medium, and high); however, this is difficult to perform in "-omics" strategies, but repeats of a complete analysis at least three times should be obligatory. There are no strict performance regulations for these procedures, but relying on just a single experiment is against the fundamental rules of "-omics" experimental design.

Inter-assay precision defines precision obtained between independent analyses performed at various occasions (eg, another day or sometimes by another operator), which is another important feature of repeatability and represents the precision obtained between different laboratories. Therefore it is extremely important to collaborate in various comparative tests, interlaboratory tests or analytical contests to independently verify one's own performance criteria. This is also beneficial in cases where participating laboratories use various approaches and instrumentation to analyze an identical sample.

An "-omics" methodology leading to the discovery of a potential biomarker should be reliable and thus sensitive and specific. This means that a set of data representing the protein profile is detected at the appropriate concentration level and is also specific for a given pathophysiology. In an ideal case, sensitivity and specificity should equal 100%. This means that the strategy is sensitive enough to detect the entire protein pattern and is also specific to identify a particular health state.

The analyte's recovery depends on the sample type, processing and also concentration, including interfering impurities in the biological matrix. The analyte's recovery can be performed using a defined amount of standard(s) applied (spiked in) at various concentrations. This method is closely related to the linearity of the calibration curve for quantitation. It is worth noting that linearity range varies for a given method of sample recovery, and therefore an analyte (sample) recovery experiment should be within the limits of linearity. To avoid problems with daily variations of recovery, an internal standard (or several) should be added to the sample before its processing. A calibration curve is helpful in estimation of detection level (sensitivity) under conditions that the sensitivity for a standard and pure substance might significantly differ from the sensitivity in a complex mixture (sometimes by few orders of magnitude). In other words, when a pure substance is being detected at an atto-molar level, a similar component in a complex biological mixture might be detected at the pico-molar level.

14.7 Validation of Performance of Instruments

As with any other analytical instrumentation, mass spectrometers are a source of errors in everyday laboratory practice and require recurrent calibrations [13]. Depending on the type of mass spectrometer, manufacturer recommendations and adopted laboratory practice, calibrations, etc. may vary from place to place. For example, MALDI-TOF instruments must be calibrated at least every day; however, many researchers calibrate them every time they analyze samples, which can be several times a day. This is quite common when multiple investigators use one instrument, switching from positive to negative ion mode or changing measuring m/z range. Based on Thermo Scientific recommendations, calibration of the Orbitrap mass spectrometer should be performed once a week, however, some laboratories calibrate it every other day. Development of mass spectrometers leading to increased mass accuracy, resolution and sensitivity makes calibration and validation of instrument performance even more important for comparisons of large data sets, in particular between laboratories.

Validation of mass spectrometers is one part of the procedure, and the other part is validation of liquid chromatography systems, which are often connected in tandem on the front end of the mass spectrometer. In most proteomic applications, nano-flow systems are used, and although technology in this area has improved tremendously in recent years, keeping a steady flow at the nano-flow level per minute remains a challenge [15]. In electrospray ionization using microcapillary columns, fluctuations in flow of the mobile phase may have a profound effect on peptide measurements. Nano-flow can be measured using capillary graduate pipettes, and although such measurement is not very precise, it is usually sufficient to achieve good spray of a mobile phase. Column batches, particularly homemade, are also a source of possible problems, eg, in case of the label-free experiments, where maintaining highly reproducible retention times over a long period is crucial for a successful experiment. Besides the above factors, there are no uniform rules of validation separation quality when thousands of peptides are eluted.

After these above steps are successfully completed, sensitivity of the system as one piece needs to be tested. A known amount of standard tryptic digest of BSA or other protein is often used. In this situation, sensitivity is usually expressed in a number of peptides identified when a certain amount of mixture is loaded. We have to accept that depending on the laboratory settings, these measures may vary. For example, in a core facility setting, sensitivity, which is also a validation point, can be arbitrarily expressed as a guarantee of high confidence identification of at least two unique high-confidence peptides when 10 fmole of standard tryptic digest of BSA is loaded. It does not mean that a nano-LC MS/MS system cannot be more sensitive, and in many cases it is, but the predetermined threshold constitutes a guarantee for core users. Such measures are easy to implement in ESI systems and are more difficult and more time-consuming to be formally employed in MALDI experiments, even if we assume that the instrument collects measurements from 1000 shots per spot and a mixture of analyte with matrix is evenly distributed throughout the target. Moreover, exploitation of the laser over a time may also significantly contribute to decreasing sensitivity and resolution of the measurements.

Validation of instruments for electrophoresis-based experiments such as 2D PAGE, or 1DE is even harder because of the central role that the polyacrylamide slab gel plays in this technique. As much as IPG strips undergo quality control at manufacturers' sites, gels are still often made in individual laboratories. Quality and reproducibility of manufactured gels improved during the last decade; nevertheless, we observe variability from batch to batch and from manufacturer to manufacturer. Another limitation is lack of clear performance criteria for analysis of complex biological samples in gel electrophoresis comprised of multiple elements such as linearity of polyacrylamide gel gradient, completeness of protein denaturation, etc., which are all inherently different. We have to keep in mind that reproducibility of separations using immobilized pH gradient (IPG) strips, whether for 2DE or OFFGEL techniques, have the same caveats.

In analytical chemistry, instrument calibrations and validation are quite well defined in many regulatory guidelines and requirements, such as Good Laboratory Practice (GLP), Good Clinical Practice (GCP) and Good Manufacturing Practice (GMP). It will take some time to transform and adapt these guidelines for the purpose of validating analytical components of proteomic profiling of highly complex biological samples.

14.8 Bioinformatics: Validation of Output of Proteomic Data

Bioinformatics offers tools to crunch the ever-expanding data from high-throughput studies and is very rich and diverse, consisting of open-source as well as licensed software packages. All of them are based on an algorithm that is sitting in a "black box" and not fully or even partially visible, known and/or understood by the users. Descriptions provided by the authors (creators, programmers) of these software packages are styled using language that is not necessarily understandable to others, in particular to those who have limited programming and statistical knowledge. This applies to those who are at the early stages of a scientific career as well as those who use proteomics as only one of the experimental approaches to test part of a hypothesis. If we accept the definitions/descriptions of software validation/verification as proposed by J.W. Ho and M.A. Charleston (http://sydney.edu.au/engineering/it/research/conversazione_2009/hoj.pdf), then software verification is a check that the algorithm is correctly implemented in the source code, meaning that the software is built right. Software validation is a check that the software performs what it is intended to perform. The end-user does not have answers to such questions and takes for granted that the software he or she is using is the right one.

Data resulting from high-throughput proteomic experiments, in which dynamic biological systems or models are tested, contain multiple variables, usually with high levels of noise and background information, a substantial number of gaps in the data, and stepwise or continuous gradient of confidence in correctness of data acquisition, sensitivity, specificity, etc. A multiplicity of factors, both intrinsic and extrinsic, may affect identification of molecules in biological systems, compartmentalization of molecules and integration of the information gained from various experiments [16].

Taken together, this poses an enormous challenge extending beyond purely analytical aspects of the problem in extracting novel information, which is not visible at first glance. Moreover, part of the data can be easily thrown out during the "data-cleaning" process. By "data-cleaning" process, we understand the ability to set filters that are provided by software packages and are set by individual investigators. Because it is difficult to grasp information in the form of large Excel files, which are the usual output files of massive mass spectrometry data, investigators use clustering techniques leading to a visual presentation, which is based on existing knowledge. The danger is that the approach to such data analysis might be highly biased by individual perspective on what an "appropriate" result should be, making it harder to accept and possibly even ignoring unexpected and contradictory data representing novel information [17].

14.9 Cross-Validation of Initial Results

While proper experimental design may dictate that one method is the best choice economically and empirically for what the investigator wishes to study, cross-validation using an orthogonal method is also needed. For instance, while a western blot may show changes within a certain protein within a biological system, that change may be less pronounced through the targeted use of mass spectrometry methods, such as MRM (multiple reaction monitoring) or SRM (selective reaction monitoring) [18]. Such technologies used to validate initial results include ELISA (enzyme-linked immunosorbent assay), western blot, immunohistochemistry, TMT-SRM (tandem mass tag selective reaction monitoring), SRM/MRM, AQUA (absolute quantification) mass spectrometry, TMT calibrator, etc.

14.10 Proteomics and Regulatory Affairs

Genetic engineering of plants and animals to insert elements protecting from insects, viral or fungal diseases is the inevitable future, and despite opposition there are no signs of slowing progress because it provides a means for more efficient food production for an ever-growing human population worldwide. At the same time and in response to a demand from the general public, governments and governmental agencies introduce new regulations and requirements. One example is the Genetically Modified Organism Compass (http://www.gmo-compass.org/eng/home/), a European source of information about genetically modified organisms from research to commercialization. One can find here extended information about plants used for consumption as well as about plants which efficiently produce valuable pharmaceuticals, biodegradable materials for industry, or enzymes that can improve animal feed, known as molecular farming or bio-pharming. While we unquestionably benefit from genetic modifications, many subsequent questions remain unanswered. For example, what if, eg, microorganisms can take up genetic material, integrate it into their genome and pass on to other organisms, such as insects, thus making them resistant to pesticides? At this point of our knowledge, the precise and direct insertion of genetic material is not available and we do not understand how random insertion affects the organism as a whole. More importantly, we do not know what global proteomic changes are made due to genetic manipulation and how these changes affect the overall balance between benefits and potential adverse effects. These questions can be addressed by performing fully unbiased proteomic profiling; however, its value exists only if such profiling can be validated.

The objectives of gene therapy are to replace a mutated gene that causes disease with a healthy copy of the gene, inactivating or "knocking out" the mutated gene that is functioning improperly. Manipulation of the human genome to accomplish these goals has multiple challenges to the extent that there is no Food and Drug Administration (FDA) approved gene therapy treatment product for sale, which makes this rather a therapy of the future. Therefore, at this point we are not asking about consequences that gene introduction may have on the overall proteome of individual cells, tissues, organs and the entire organism. If the malfunctioning gene is not replaced at the exact location and the newly introduced gene has its own regulatory elements for expression, the "proteomic consequences" might not be predictable. We can foresee that when gene therapy products are eventually available as prescription therapy, determinations of consequences at the protein level will gain importance. At the same time, a question of validation of full unbiased proteomics profiling will become of increasing importance. This will be followed by increasing pressure of regulatory agencies to establish, although initially preliminary, a set of standards of accuracy, precision, sensitivity and specificity of proteomic profiling with quite rigorous quality control and quality assurance. As much as this issue may seem to be in the rather distant future, rapid technological developments shown during the last two decades may make it urgent reality sooner than expected.

As the field of proteomics matures along with computer-assisted and automated technologies, proteomics-based discoveries appear as an attractive and profitable undertaking. However, protection of intellectual property in proteomics has its own set of legal challenges. As much as it is possible to invent worldwide, patenting new discoveries worldwide is not always economical or affordable. We can expect that with further developments in proteomics, legal rules will mature and will enable investigators to protect their discoveries and profits associated with intellectual and financial investment in science. We also refer readers to an article "Patentability and Maximum Protection of Intellectual Property in Proteomics and Genomics" by R.J. Warburg, A. Wellman, T. Buck, and A.E. Ligler Schoenhard published in 22 Biotechnology Law Report 264 Number 3, June 2003.

References

[1] Kuehl RO. Design of experiments: statistical principles of research design and analysis. 2nd ed. Pacific Grove: CA Duxbury Press; 1999.

[2] Lee JK, Cui X. Experimental designs on high-throughput biological experiments. Whiley; 2010.

[3] Simon RM, Korn EL, McShane LM, Radmacher MD, Wright GW, Zhao Y. Design and analysis of DNA microarray investigations (Statistics for biology and health). Springer; 2004. 209 p.

[4] Altelaar AM, Heck AJ. Trends in ultrasensitive proteomics. Curr Opin Chem Biol 2012;16(1–2):206–13. Epub 2012/01/10.

[5] Idikio HA. Immunohistochemistry in diagnostic surgical pathology: contributions of protein life-cycle, use of evidence-based methods and data normalization on interpretation of immunohistochemical stains. Int J Clin Exp Pathol 2009;3(2):169–76. Epub 2010/02/04.

[6] Goel S, Duda DG, Xu L, Munn LL, Boucher Y, Fukumura D, et al. Normalization of the vasculature for treatment of cancer and other diseases. Physiol Rev 2011;91(3):1071–121. Epub 2011/07/12.

[7] Vogel JS, Giacomo JA, Schulze-Konig T, Keck BD, Lohstroh P, Dueker S. Accelerator mass spectrometry best practices for accuracy and precision in bioanalytical (14)C measurements. Bioanalysis 2010;2(3):455–68. Epub 2010/11/19.

[8] Glanzer JG, Enose Y, Wang T, Kadiu I, Gong N, Rozek W, et al. Genomic and proteomic microglial profiling: pathways for neuroprotective inflammatory responses following nerve fragment clearance and activation. J Neurochem 2007;102(3):627–45. Epub 2007/04/20.

[9] Enose Y, Destache CJ, Mack AL, Anderson JR, Ullrich F, Ciborowski PS, et al. Proteomic fingerprints distinguish microglia, bone marrow, and spleen macrophage populations. Glia 2005;51(3):161–72. Epub 2005/03/30.

[10] Pottiez G, Jagadish T, Yu F, Letendre S, Ellis R, Duarte NA, et al. Plasma proteomic profiling in HIV-1 infected methamphetamine abusers. PLoS One 2012;7(2):e31031. Epub 2012/02/24.

[11] Leitner A, Sturm M, Lindner W. Tools for analyzing the phosphoproteome and other phosphorylated biomolecules: a review. Anal Chim Acta 2011;703(1):19–30. Epub 2011/08/17.

[12] Brewis IA, Brennan P. Proteomics technologies for the global identification and quantification of proteins. Adv Protein Chem Struct Biol 2010;80:1–44. Epub 2010/11/27.

[13] Gil J, Cabrales A, Reyes O, Morera V, Betancourt L, Sanchez A, et al. Development and validation of a bioanalytical LC-MS method for the quantification of GHRP-6 in human plasma. J Pharm Biomed Anal 2012;60:19–25. Epub 2011/12/14.

[14] Geiger T, Cox J, Ostasiewicz P, Wisniewski JR, Mann M. Super-SILAC mix for quantitative proteomics of human tumor tissue. Nat Methods 2010;7(5):383–5. Epub 2010/04/07.

[15] Donato P, Cacciola F, Tranchida PQ, Dugo P, Mondello L. Mass spectrometry detection in comprehensive liquid chromatography: basic concepts, instrumental aspects, applications and trends. Mass Spectrom Rev 2012;31(5):523–59. Epub 2012/03/03.

[16] Fernie AR, Stitt M. On the discordance of metabolomics with proteomics and transcriptomics: coping with increasing complexity in logic, chemistry, and network interactions scientific correspondence. Plant Physiol 2012;158(3):1139−45. Epub 2012/01/19.

[17] Kell DB, Oliver SG. Here is the evidence, now what is the hypothesis? the complementary roles of inductive and hypothesis-driven science in the post-genomic era. BioEssays 2004;26(1):99−105. Epub 2003/12/30.

[18] Aebersold R, Burlingame AL, Bradshaw RA. Western blots versus selected reaction monitoring assays: time to turn the tables? Mol Cell Proteomics September 2013;12(9):2381−2. http://dx.doi.org/10.1074/mcp.E113.031658. Epub 2013/06/10.

INDEX

'*Note:* Page numbers followed by "f" indicate figures and "t" indicate tables.'

A

Absolute quantitation, 146–147, 151–152
  AQUA, 147
  Protein Standard for Absolute Quantification (PSAQ™), 147
  QconCAT, 147
Accuracy validation, 256–258, 257f
Acetone, for protein precipitation, 35
Acetonitrile, 58–60
Acetylation, 15
Acrylamide concentration correlation with separated species molecular weight, 118t
S-adenosy-L-Lmethionine (SAM), 18–19
Agarose, 70, 83
Agarose gel electrophoresis, 119–120, 120f
  data storage, 127
  sample preparation, 120
  separation conditions, 120–121
Albumin and IgG Removal Kit, 102–103
Albuminome, 107–113
Amino acid sequences, 21f
  and separating conditions, 15–17
Amino acid-containing peptides labeling, 153–154
Amino acids
  charge on, 8
  containing sulfur, 17–19
  hydrophilicity, 8–10
  hydrophobicity, 8–10
    surface area and, 10
  properties of, 8
4-(2-Aminoethyl)benzenesulfonyl fluoride (AEBSF), 27–28
Ammonium persulfate (APS), 83

Ammonium sulfate, 55–56
Ampholytes, 82
Analysis of variance method (ANOVA), 228–229
Analyte's recovery, 263
Analytical chemistry, 4–5
  importance of, 1–2
Analytical validation, 239–240
  aberrant results and errors, 247–249
  accuracy determination, 244–245
  calibration curve, 245–246
  further development, 249
  linearity, 245–246
  liquid chromatographic methods, 241–244
  precision determination, 245
  recovery determination, 244
  selectivity determination, 246–247
  sensitivity determination, 245–246
  specificity determination, 246–247
  stability determination, 247
Anion exchangers, 68–69, 71–73, 182–183
Antipain, 27–28
Aprotinin, 27–28
AQUA, 147
Aurum Serum Protein minikit, 102–103

B

16-BAC/SDS-PAGE, 135–136
Balanced incomplete block design, 220
Ballast proteins, 57–58
Bayesian methods, 229–230
Benjamini Hochberg (BH) method, 231–232
Benzyldimethyl-n-hexadecylammonium chloride (16-BAC), 135–136

Beta-glycerophosphate, 28
Bioinformatics, 1–2, 266–267
Biological replicate, 215–216
Biomolecules
  amino acid sequence, and separating
    conditions, 15–17
  amino acids
    containing sulfur, 17–19
    hydrophilicity, 9–10
    hydrophobicity, 9–10
    cysteine, 17–19
    methionine, 17–19
  peptides, major features and
    characteristics of, 8
  posttranslational modifications,
    effect of, 15
  protein folding, 21–22
    and protein–protein interactions,
    22
  proteins
    fragmentation, effect of, 11–15
    identification and characterization,
    19
    major features and characteristics
    of, 8
    moonlighting phenomenon, 20,
    22–23
    structure–function relationship,
    20–21
Bio-Rad Proteominer™ methodology, 40f
BioWorks™, 201–202
Bisacrylamide, 121–122
Blue native electrophoresis (BNE),
  122–123
Blue Sepharose 6 Fast Flow affinity
  chromatography, 103

C
Capillary columns, 86
  column diameter versus flow rate
    used, 87t
  column selection, 95t, 97t–98t
  conventional columns, 87–88, 88f
  monoliths, 89–94
  recent developments, 96–98
Capillary isoelectric focusing (CIEF),
  83–84. *See also* Isoelectric
  focusing (IEF)
Cation exchangers, 68–69, 71–73,
  182–183

Cell culture, 31–32
Cell fractions, 31–32
Cell sorting, 31–32
Cells isolated from blood, 32t
Cerebrospinal fluid protein extraction,
  36–37
  immunodepletion techniques,
    104–105
Chemical labeling, 213–214
Chloroform/methanol extractions,
  52–53
Chromatofocusing, 182–183
Chromatographic methods, of protein
  fractionation, 38–39
  interfering and unnecessary proteins,
    39
  salt removal, 38–39
Chymostatin, 27–28
Clathrate structure, 9–10
Clear native electrophoresis (CNE),
  122–123
Cluster analysis, 232
Collision-induced dissociation (CID),
  151–152
Continuous-bed chromatography,
  93–94
Continuous-density gradient, 34
Coomassie Brilliant Blue (CBB),
  122–123, 126, 134, 134t
Cross-linked polystyrene-
  divinylbenzene polymer,, 70f
Cross-validation, 267. *See also* Validation
  in proteomics
Crude protein extract, 34–36
  fractionation based on size-exclusion
    filters, 37–38
  protein precipitation,
    35–36
C-terminal labeling, 152–153
  advantages and drawbacks,
    148t
Cultures
  of established cell lines, 32t
  of primary cells, 32t
Cyanine dyes, 140–141, 141t
α-Cyano-4-hydroxycinnamic acid
  (CCA), 184–185
Cysteine, 17–19
Cytoselective method of cell separation,
  31–32

D

Data output validation, 266–267
Data preprocessing, for statistical
analysis, 224–227
data preparation and filtering,
224–225
missing value imputation,
226–227
normalization, 225–226
transformation, 225
Data-cleaning process, 266–267
Data-dependent acquisition (DDA)
analysis, 162–164, 171–172
Data-independent acquisition (DIA)
analysis, 162, 165, 171–172
Denaturation, 131–132
Density gradient centrifugation, 31–34
Density markers, 34
Deoxycholate (DOC), 60–61
Detergents, 44–46
"cloud point" (CP), 45–46
CMC (critical micelle concentration),
44–45
removal, 46
types, 45
Dextrans, 70
protein precipitation, 56
Difference gel electrophoresis (DIGE),
126, 139, 140f
cyanine dyes, 140–141, 141t
fluorescent dyes, 140
internal standard, 141–142
pros and cons, 142
Differentially expressed proteins,
228–230
2,5-Dihydroxybenzoic acid (DHB),
184–185
3,5-Dimethoxy-4-hydroxycinnamic acid
(sinapinic acid or SA), 184–185
Directed acyclic graphs (DAGs), 234
Disulfide bridges, 18–19
Dithiothreitol (DTT), 131–132
DNA removal, 46–48

E

Electrophoresis, 117. *See also* Gel
electrophoresis; Gel
electrophoresis, two-dimensional
Electrospray ionization (ESI),
184–185, 186f–187f, 204

Elutriation technique, 31–32
"Equalizing" method, 39, 40f
Ethanol/methanol precipitation, 58
Ethylene-bridged hybrid (BEH) particle
structure, 240–241
Ethylenediaminotetraacetic acid
(EDTA), 27–28
Experimental design, 214–224
biological replicate, 215–216
experimental layout, 216–217
label assignment, 216–217
label-free experiment layout, 217
randomization, 215
sample size calculation, 221–224
stable isotope labeling experiment
layout, 219–221
balanced incomplete block design,
220
label swapping, 219
Latin square design, 219–220
loop/cyclic design, 221
reference design, 220–221
technical replicate, 215–216
and validation, 258–259
Experimental layout, 216–217
Extracted ion chromatogram (XIC),
168

F

False discovery rate (FDR), 231
Family-wise error rate (FWER), 231
Flow cytometry, 31
Fluorescent staining, 126, 134–135, 134t
in difference gel electrophoresis,
140–141
Fourier Transform Ion Cyclotron
Resonance MS (FT-ICR-MS),
185–187
Fourier Transform Orbitrap mass
spectrometry (FT-Orbitrap-MS),
188

G

Gel electrophoresis, 117–119
advantages and disadvantages, 125t
agarose gel electrophoresis, 119–120,
120f
sample preparation, 120
separation conditions, 120–121
conditions, 119

Gel electrophoresis (*Continued*)
   data storage, 127
   difference gel electrophoresis (DIGE), 126
   isotope labeling, 127
   obstacles during, 125t
   polyacrylamide gel electrophoresis, 121–123, 122f
   sodium dodecyl sulfate-polyacrylamide acrylamide electrophoresis (SDS-PAGE), 123–124
   sample preparation, 124
   staining techniques, 124–126
   fluorescent staining, 126
Gel electrophoresis, two-dimensional, 128–129
   advantages, 135
   difference gel electrophoresis (DIGE), 139, 140f
   disadvantages, 135–136
   gel staining, 133–135, 134t
   isoelectric point, 129–130, 130f
   molecular weight, 131–133, 132f
   protein quantitation, 136–139, 137f
Gel-based quantitative proteomics, 149
   advantages and drawbacks, 148t.
   See also Difference gel electrophoresis (DIGE)
Gel-free quantitative proteomics, 149–150, 150f
Genetically Modified Organism Compass, 267–268
Glass beads shaking/crushing, homogenization method, 30t
Global Proteome Machine, 203
Glycidyl methacrylate (GMA), 92
Glycoprotein analysis, 58
Glycosylation, 42–44
Graphical Gaussian models (GGMs), 234

H
HeavyPeptide, 201–202
Hierarchical clustering, 232
High-performance liquid chromatography (HPLC), 87–88, 96, 240–241, 242f
   versus ultra-performance LC (UPLC), 89t

High-resolution clear electrophoresis (hrCNE), 122–123
Homogenization, 28–29, 30t
   common methods, 30t
   control by internal standard addition, 29
   and isolation of organelles, 31–34
Hydrogel, 58–60
Hydropathy index, 9
   for protein, 9
Hydrophilic interaction liquid chromatography, 179
Hydrophilicity, 8–10
Hydrophobic interaction chromatography, 182
Hydrophobic proteins, 52–53, 135–136
Hydrophobicity, 8–10

I
Image warping, 137–138
Immobilized polyacrylamide gel (IPG) strips, 83, 129–130, 133
Immunodepletion techniques, 57–58, 102–104, 113
   albuminome, 107–113
   capacity of columns and other devices, 104–105
   human plasma proteins, 109t–112t
   quality control, 106–107
   reproducibility, 105–106, 107f
Inline LC-MS top-down proteomics, 177–180
   hydrophilic interaction liquid chromatography, 179
   reversed-phase liquid chromatography (RPLC), 177–179
   size-exclusion chromatography (SEC), 179–180, 181f
Inline separation, 176
Instruments performance validation, 263–266
Inter-assay precision, 262–263
International Protein Index (IPI), 198–199
Intra-assay precision, 262
Ion exchange chromatography, 182–183
   buffers in, 73–74, 74t
   choice of ion exchanger, 71–73

choice of strong or weak ion
exchanger, 73
historical perspective, 66
monolithic columns, 70–71
principle of, 66–68, 68f
in proteomic studies, 74–76
types, 68–71, 69t
Ionic detergents, 45
Ionization techniques, 184–185
electrospray ionization (ESI),
184–185, 186f–187f, 204
matrix-assisted laser desorption/
ionization (MALDI), 184–185, 204
Isobaric tags for relative and absolute
quantification (iTRAQ), 151–152,
151f, 213–214
advantages and drawbacks, 148t
Isoelectric focusing (IEF), 8, 85,
129–130, 130f
capillary IEF (CIEF), 83–84
drawback, 80
in liquid state, 82
in living organisms, 84
principles of, 77–80, 78f–79f
sample preparation, 80–81
using immobilized polyacrylamide gel
(IPG) strips, 83
Isoelectric point (pI), 71
precipitation, 56–57
Isolated cell populations, 31–32
advantages and limitations of using,
32t
Isotope labeling, 127
Isotope-coded affinity tagging (ICAT),
149–150, 150f, 213–214
advantages and drawbacks, 148t
Isotopically differentiated 2-
nitrobenzenesulfenyl chloride
(NBSCI), 153–154
advantages and drawbacks, 148t

K

Kurskal–Wallis test, 228–229
Kyte–Doolittle analysis, 10

L

Label assignment, 216–217
Label swapping, 219
Label-free techniques, 155–157, 214

advantages and drawbacks, 148t
experiment layout, 217
SWATH-MS data acquisition, 165–168
Large-molecular-weight molecules, 117
Laser capture microdissection, 32
Laser-induced fluorescence microscopy
or mass spectrometry, 96
Latin square design, 219–220
Leupeptin, 27–28
Linear model, 228–229
Lipids removal, 46–48
Liquid chromatographic methods,
240–241
Liquid chromatography coupled to mass
spectrometry (LC/MS), 240–241
Liquid chromatography, 8, 86
column selection, 97t–98t
HPLC versus UPLC, 87–88, 89t
Liquid nitrogen crushing,
homogenization method, 30t
Loop/cyclic design, 221
Lower limit of quantitation (LLOQ),
244–246
Lysis buffers, homogenization method,
30t

M

Mann–Whitney test, 228–229
MarkerView Software, 169
Mass accuracy, of precursor ion, 195
Mass spectrometry
-based quantitative proteomics,
213–214
label-free quantification, 214
stable isotope labeling, 213–214
data searches, 203–205
miscleavage, 205
search parameters, mass tolerance,
204–205. See also Post-database
search data processing
of intact proteins, 183–188
Mass-coded abundance tagging
(MCAT), 153–154
advantages and drawbacks, 148t
Matrix-assisted laser desorption/
ionization (MALDI), 184–185, 204
MaxQuant, 202–203
Mechanical, rotor-stator,
homogenization method, 30t

Metabolic labeling, 154–155, 155f,
213–214
  advantages and drawbacks, 148t
Methacrylate capillary monoliths, 92
Methacrylate-based monoliths, 92
Methionine, 17–19
Micro-total analysis systems
(microTAS), 96
Migration ratio, 117
Molecular separation range as a function
of agarose gel concentration,
118t
Molecular Weight Search (MOWSE),
200–201
  MOWSE II version, 200–201
  MOWSE III version, 200–201
Molecular weight, 131–133, 132f
Monoliths, 89–95, 90f
  methacrylate-based monoliths, 92
  organic-based monoliths, 91–92, 91f,
  93t
  silica-based monoliths, 89–91, 93t,
  94–95
  styrene-based monoliths, 92–94
Moonlighting phenomenon, 22–23
(2-(N-Morpholino)ethanesulfonic acid
(MES), 82
(3-(N-Morpholino)propanesulfonic acid
(MPOS), 82
MS-Align+, 188–189
MS-Deconv, 188–189
Mucins, 42–43
Multiple Affinity Removal System,
102–103
Multiple comparisons across proteins,
230–232
  false discovery rate (FDR), 231
  family-wise error rate (FWER), 231
Multiple reaction monitoring (MRM),
267

N
Nanoelectrospray emitters, 90–91, 91f
National Center for Biotechnology
Information Nonredundant
Database, 198
Net charge of a protein, 71–72, 72f
Nonionic detergents, 45
Nonporous silica (NPS), 177–178

N-terminal isotope-encoded tagging
(NIT), 150–151
  advantages and drawbacks, 148t
N-terminal labeling, 150–152
  advantages and drawbacks, 148t

O
Offline fractionation, 65
Offline LC-MS top-down proteomics,
180–183
  hydrophobic interaction
  chromatography, 182
  ion-exchange chromatography,
  182–183
Offline separation, 176
Oligosaccharide chains, 42–43
"-Omics" methodology, 263
Online fractionation, 65
OpenSWATH software, 171
Orbitrap technology, 188
  combined with an ion trap, 195
Organic solvent, 178–179
Organic solvent-driven precipitation,
57–60
Organic-based monoliths, 91–92, 93t

P
PeakView software, 168
Pepsin digestion of IGF2, peptides
generated by, 11–15, 12t–14t
Pepstatin A, 27–28
Peptide Atlas, 167
Peptides, 11
  chromatographic separation,
  15–17
  gel-based separation, 17
  major features and characteristics of, 8
  physicochemical properties, 12t–14t
  posttranslational modifications, effect
  of, 15
  purification of, 40–44
  sequencing, by mass spectrometry, 10
Percoll™, centrifugation in, 31–32
Phenol extraction, 52–53
Phenylmethanesulfonyl fluoride
(PMSF), 27–28
Phosphopeptides, 41–42
Phosphorylation, 41–42, 43f
PhosphoSitePlus (PSP), 199

Planetary discs blending, homogenization method, 30t
Plug-and-play LC-MS, 96–97
Plug-and-use fritting technology, 96–97
Polar solvent, 178–179
Polyacrylamide gel electrophoresis, native, 121–123, 122f
  data storage, 127
Polyethylene, protein precipitation, 56
Polysaccharide, 119–120
POROS™ Affinity Depletion cartridges (Anti-HSA and Protein G), 102–103
Porous-layer open-tubular (PLOT) columns, 93–94
Post-database search data processing, 205–207
  ProteoIQ, 207
  Scaffold software, 206
  Skyline software, 207
  Trans Proteomics Pipeline, 206
Posttranslational modifications (PTM), 15, 41, 84, 188–189
  searches for, 207–208
Potter-type (PTFE-glass or PTFE–PTFE crushers, homogenization method, 30t
Precision validation, 256–258, 257f
Pressure, homogenization method, 30t
Principal component analysis (PCA), 169, 233
PROC itraqnorm, 225–226
Pro-Q dyes, 126
ProSightPC 3.0, 188–189
Protease inhibitors, 26–28
Protein databases, 196–199
  Human Genome and Protein Database (HGPD)
  International Protein Index (IPI), 198–199
  NCBI database, 198
  PhosphoSitePlus (PSP), 199
  redundancy in, 196
  Swiss-Prot, 197
  UniProt database, 196–197
  UniRef, 197–198. See also Post-database search data processing
Protein Expression Assembler, 169

Protein expression data, statistical analysis, 227–234
  cluster analysis, 232
  differentially expressed proteins, 228–230
  multiple comparisons across proteins, 230–232
    false discovery rate (FDR), 231
    family-wise error rate (FWER), 231
  principal component analysis (PCA), 233
  protein networks, 233–234
  time-dependent proteins, 230
Protein Expression Workflow, 169
Protein extraction, 51–52
  hydrophobic protein, 52–53
  protein precipitation. See Protein precipitation
  protein solvation, 53–55
  salting out process, 55–56
Protein networks, 233–234
Protein precipitation, 35–36, 51–55, 54f
  isoelectric point precipitation, 56–57
  organic solvent-driven precipitation, 57–60
  trichloroacetic acid (TCA) precipitation, 60–61
Protein Standard for Absolute Quantification (PSAQ™), 147
ProteinPilot software, 166–167, 202
Proteins, 8
  characterization, 19
  folding, 21–22
    and protein–protein interactions, 22
  fractionation of, 8
  fragmentation, effect of, 11–15
  hydropathy index for, 9
  identification, 11–15, 19
  major features and characteristics of, 8
  moonlighting phenomenon, 20, 22–23
  posttranslational modifications, effect of, 15
  reduction and alkylation of, 132f
ProteoIQ, 207
Proteome Discoverer™, 201–202
Proteome map, 138

Proteomic sample preparation, 26
  cerebrospinal fluid protein extraction,
    36–37
  crude protein extract, 34–36
    chromatographic methods of
      protein fractionation, 38–39
    fractionation based on size-
      exclusion filters, 37–38
  detergents removal, 44–46
  DNA removal, 46–48
  homogenization, 28–29
    common methods, 30t
    and isolation of organelles, 31–34
  lipids removal, 46–48
  peptide purification, 40–44
    glycopeptides, 42–44
    phosphopeptides, 41–42
  protease inhibitors, 26–28
  serum/plasma protein extraction,
    36–37
Proteomics, 1, 2f, 4–5
  bioinformatics analyses, 1–2
  biologist's approach, 3
  chemist's approach, 3
  design considerations, 52. *See also*
    Experimental design

Q
QconCAT, 147
Q-TOF-MS instruments, 188
Quantitation, 146, 157–158
  absolute quantitation, 146–147, 151–152
  amino acid-containing peptides
    labeling, 153–154
  C-terminal labeling, 152–153
  gel-based quantitative proteomics,
    148t, 149
  gel-free quantitative proteomics,
    149–150, 150f
  label-free techniques, 155–157
  metabolic labeling, 154–155, 155f
  N-terminal labeling, 150–152
  proteolytic ^{18}O labeling, 152–153, 153f
  tandem mass spectrometry, 162–165

R
Randomization, 215
Randomized complete block designs
  (RCBD), 217, 218f

Reference design, 220–221
Regulatory affairs, 267–269
Relative quantitation
  amino acid-containing peptides
    labeling, 153–154
  C-terminal labeling, 152–153
  gel-based quantitative proteomics,
    148t, 149
  gel-free quantitative proteomics,
    149–150, 150f
  label-free techniques, 155–157
  metabolic labeling, 154–155, 155f
  N-terminal labeling, 150–152
  proteolytic ^{18}O labeling, 152–153, 153f
Relevance networks (RNs), 234
Replication
  biological replicate, 215–216
  technical replicate, 215–216
Reproducibility, 105–106, 107f
  validation, 262–263
Resin, 69–70
Reversed-phase liquid chromatography
  (RPLC), 177–179
Reverse-phase chromatography (RPC),
  9–10
Riboflavin, 83

S
Salting out process, 55–56
Sample loss validation, 262–263
Sample size calculation, 221–224
SAS/STAT® software, 225–226
Scaffold software, 206
Search engines, 199–203, 200f
  Global Proteome Machine, 203
  MaxQuant, 202–203
  Molecular Weight Search (MOWSE),
    200–201
  ProteinPilot, 202
  Proteome Discoverer™, 201–202
  Sequest algorithm, 201–202
  SpectraST, 203
  X! Tandem, 203
Sec translocase system, 21–22
Selective reaction monitoring (SRM),
  267
Sephadex™, 70, 153–154
Sepharose™, 70
Seppro SuperMix System, 103

Seppro™ IgY14, 103–105
Sequential Windows Acquisition of All
    Theoretical Spectra-Mass
    Spectrometry (SWATH-MS) data
    acquisition, 162, 166f, 171–172
  overview, 168–171
    MarkerView software, 169
    OpenSWATH software, 171
    Skyline software, 170–171
    Spectronaut software, 170
    z-transformation, statistical analysis
      using, 169–170
  spectral library, 166–167
  targeted data extraction, 168
  workflow, 163f
Sequest algorithm, 201–202
Serum/plasma protein extraction, 36–37
Signal-to-noise ratio (S/N), 246
Silica-based monoliths, 89–91, 93t,
    94–95
Silver staining, 126, 134, 134t
Single cell suspension of primary cells,
    32t
Size-exclusion (cut-off) filters, 37–38
Size-exclusion chromatography (SEC),
    179–180, 181f
Skyline software, 207
Sodium deoxycholate, 35–36
Sodium dodecyl sulfate (SDS), 123–124,
    131–132
Sodium dodecyl sulfate-polyacrylamide
    acrylamide electrophoresis (SDS-
    PAGE), 103, 106, 123–124
  sample preparation
Sodium fluoride, 28
Sodium orthovanadate, 28
Sodium pyrophosphate, 28
Soft ionization techniques, 183
Solid-phase extraction (SPE), 38–39,
    58–60
Sonication, homogenization method,
    30t
Spectral library, 166–167
SpectraST, 203
Spot quantities, normalization of, 138
Stable isotope labeling experiment
    layout, 219–221
  balanced incomplete block design,
    220

label swapping, 219
Latin square design, 219–220
loop/cyclic design, 221
reference design, 220–221
Stable isotope labeling with amino acids
    in cell culture (SILAC), 154–155,
    155f, 213–214
  advantages and drawbacks, 148t
Stable isotope labeling, 213–214
Statistical analysis
  data preprocessing for, 224–227
  of protein expression data, 227–234
Stepwise gradient, 33
Strong anion exchangers, 68–69
Strong cation exchangers (SCX), 38–39,
    67–69, 73, 75–76
Strong ion exchangers, 73
Structure–function relationship,
    significance in systems biology
    function, 20–21
Styrene-based monoliths, 92–94
Swiss-Prot, 197
SYPRO dyes, 126
SYPRO Ruby, 134–135

T
Tandem mass spectrometry, for
    quantitative proteomics,
    162–165, 164f
  data-dependent acquisition (DDA),
    163–164
  data-independent acquisition (DIA),
    165
Technical replicate, 215–216
Tetramethylenediamine (TEMED), 83
Time-dependent proteins, 230
Time-of-flight (TOF) instruments, 188
Top-down proteomics, 175–176
  data analysis software, 188–190
  mass spectrometry of intact proteins,
    183–188
    ionization techniques, 184–185
    mass spectrometry instruments,
    185–188
  protein separation methods, 176–183
    inline LC-MS top-down proteomics,
    177–180
    offline LC-MS top-down
    proteomics, 180–183

Trans Proteomics Pipeline, 206
TrEMBL, 197
Trichloroacetic acid (TCA)
   and ethanol mixture, protein
      precipitation, 35–36
   protein precipitation, 60–61
   and sodium deoxycholate mixture,
      protein precipitation, 35–36
Trifluoroacetic acid (TFA), 93–94
Trypsin, miscleavage, 205
Trypsin digestion of IGF2, peptides
      generated by, 11–15, 12t–14t
Type I error, 222–223, 230–231
Type II error, 230–231

U
Ultra-high-pressure liquid
      chromatography (UHPLC),
      240–241, 242f
Ultra-performance liquid
      chromatography (UPLC),
      87–88
   versus high-performance LC (HPLC),
      89t
UniProt database, 196–197
UniProtKB/Swiss-Prot, 197

UniRef, 197–198
Upper limit of quantitation (ULOQ),
      245–246

V
Validation in proteomics, 253–256, 256f
   accuracy and precision, 256–258, 257f
   cross-validation, 267
   data output validation, 266–267
   detection levels validation, 260–262
   experimental design and, 258–259
   instruments performance validation,
      263–266
   methods validation, 259–260
   reproducibility and sample loss
      validation, 262–263

W
Weak ion exchangers, 68–69, 82

X
X! Tandem, 203

Z
z-score transformation, 169–170
Zwitterionic detergents, 45